Pathways

to Mathematical Understanding

Early Childhood to Middle School

A Guide for Parents and Teachers to Develop Mathematical Thinking in Young Children

Daniel Zaccaro

Edward Zaccaro

HICKORY GROVE PRESS

Enriching the Curriculum

About the Authors

Dan lives in Dubuque, Iowa with his two children. He is a licensed School Psychologist and works as a School Improvement Facilitator for Keystone Area Education Agency. After graduating from Luther College in 2009, he earned three graduate degrees from the University of Northern Iowa: Masters of Arts in Education (MAE), Education Specialist (Ed.S), and a PK-12 Principal/Special Education Supervisor endorsement.

Ed lives outside of Dubuque, Iowa, with his wife Sara. He has been involved in various areas of education since graduating from Oberlin College in 1974. Ed holds a Masters degree in gifted education from the University of Northern Iowa and has presented at state and national conferences in the areas of mathematics and gifted education.

Hickory Grove Press 3151 Treeco Lane Bellevue, Iowa 52031

E-mail: challengemath@aol.com
Website: www.challengemath.com

Library of Congress Control Number: 2021910429
ISBN 10: 0-9854725-4-5
ISBN 13: 978-0-9854725-4-2

Books By Edward Zaccaro

- *Primary Grade Challenge Math*
- *Upper Elementary Challenge Math*
- *Challenge Math*
- *Real World Algebra*
- *Becoming a Problem Solving Genius*
- *The 10 Things All Future Mathematicians and Scientists Must Know (But are Rarely Taught)*
- *25 Real Life Math Investigations That Will Astound Teachers and Students*
- *When Math Matters*
- *Scammed By Statistics*
- *Now You Know Volume 1*
- *Now You Know Volume 2*

Table of Contents

Introduction

Why this book?

Parents know they should read to children to help develop their literacy skills. Unfortunately, when it comes to developing mathematical thinking, these same parents feel confused about how to jump-start a child's understanding of numbers.

Most math books seem to have one audience in mind: experienced teachers. They are full of terms like "subitize" or "quantitative discrimination," and often have activities that only a large class can do. ***Pathways to Mathematical Understanding*** offers a way to cut through the noise and gradually build a child's understanding of numbers and mathematical thinking.

Some ideas are limited by age and brain development, but this book contains visuals and examples to help children learn new concepts well before they are exposed to them in the classroom.

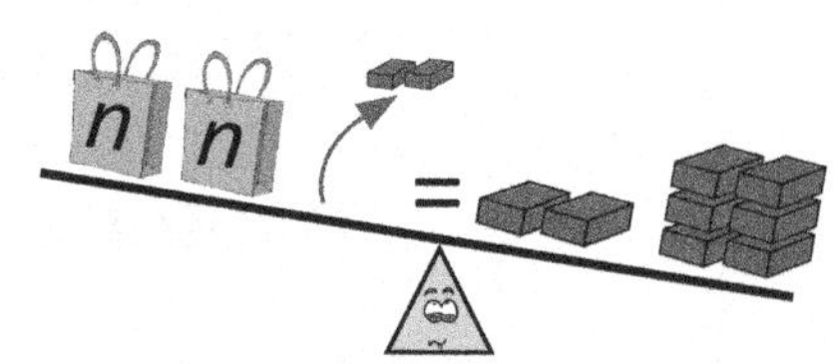

(Algebra)

Thinking over Tricks

If you spend time in American classrooms today, you are sure to hear rules that either stop working in certain grades or are completely irrelevant to math. These rules or "tricks" sometimes lead to the right answer, but limit how students ***think*** about numbers. In general, teachers are qualified and good at their jobs, but some haven't been given the opportunity to learn better ways to approach numbers with children. When students hear "knock on your neighbor's door to borrow a cup of sugar" when subtracting, they are not learning math—they are learning a series of steps that *resemble* math. Too much of this type of instruction can lead to a dependence on rote memorization, which could handicap a child's mathematical future.

The importance of learning rote math skills alongside mathematical thinking is very much like the teamwork of bricks and mortar. When separated, bricks or mortar can't build a very strong wall. However, when used together they are incredibly strong.

Look how this concept applies to reading. Picture a second grader with good decoding skills reading the following two passages.

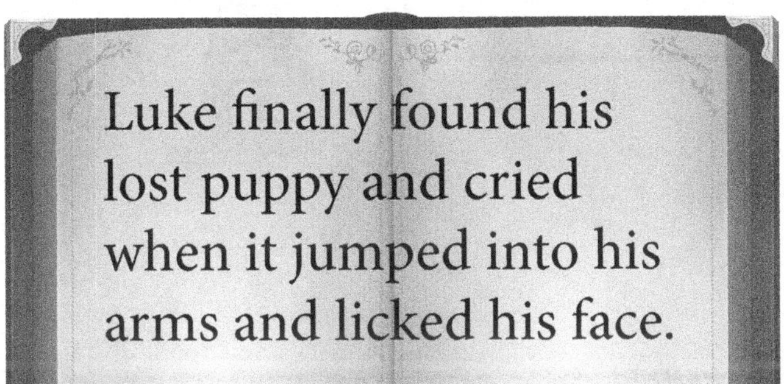

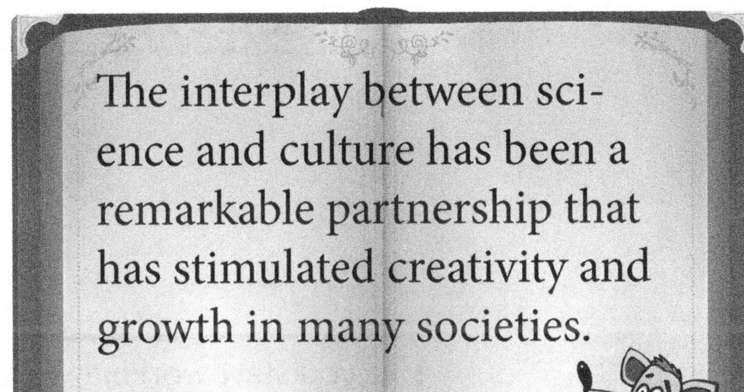

Someone with good decoding skills can "read" both passages. The first passage is not only easy to decode, but is also more likely to be understood. While the same person might have the rote skills to *decode* the second passage, he most likely will have no idea what the words mean. This is a very important concept in literacy *and* mathematics.

Look at the following problem that a student solved incorrectly because of a failure to think mathematically. The last step of solving a problem should always be to double-check your answer to see if it makes sense — **turn on the mathematical thinking part of the brain!**

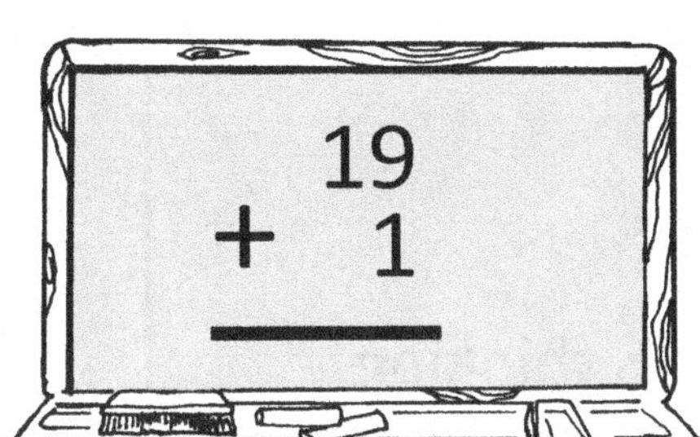

$$\begin{array}{r} 19 \\ +\ \ 1 \\ \hline 110 \end{array}$$

Let's see, I add 9 and 1 and get 10. Put the zero down and carry the 1 and get 11. The answer is 110. I am awesome! (...Right?)

A strong number sense will follow children for their entire lives and open many careers for them. These skills also help with mental math, money, time, fractions, computation, estimation, and many other number-related activities.

For example, children who understand the inner workings of numbers won't need to memorize 100 different multiplication facts. Instead, they can quickly use what they *know* to find answers for what they *don't*...

$$\begin{array}{r} 8 \\ \times\ 6 \\ \hline \end{array}$$

8 x 6... Hmm I don't remember my 8 x 6 fact, but I do know that 8 x 3 is 24. If 3 groups of 8 are 24, that means 6 groups will be twice as much, so 8 x 6 = 48.

As you start working with your child on math literacy, please keep in mind that math is often taught as all scales and no music. This book has the music! It will show parents and teachers how to explore the exciting, real parts of mathematics.

As you work to build a successful relationship between children and the educational process, remember these points:

- Passion for a subject often leads to excellence.
- Challenge and frustration are a normal part of the learning process—and life.
- Nurture your child's love of knowledge.
- Teach for true understanding, not rote steps.
- Provide enough challenge to stretch a child's thinking.

In this book

This book contains easy-to-understand explanations of important number concepts, along with activities and resources to develop them.

Section I Provides essential learning for adults to help their math conversations with children. Which questions to ask, how to build math routines, and how to foster the best environment for learning.

From this point on, each chapter includes 20-40 questions that get more difficult with each level:

Level 1:	Easy
Level 2:	Slightly challenging
Level 3:	Challenging
Genius Level:	Very challenging

Section II Introduces "Number Sense" and outlines the steps to develop it. These follow a predictable pattern as a child's brain develops.

Section III Applies these number sense skills into operating on numbers (addition, subtraction, multiplication, and division). Three important types of numbers are explained, as well as the relationship between them: fractions, percents, and decimals. This section also includes chapters on the best ways to group and prioritize math facts.

Section IV Introduces advanced topics like algebra, equations, and ratios. Negative numbers are visualized and explained.

Section V Answers for problems in **Section III** and **Section IV**

Section VI Resources to copy and use (number lines, place value cards, bingo board, etc).

Some topics—like measurement and geometry—are not covered in this book, as we prioritized only the most important number concepts. These are the foundational ideas that will support every math-related task in a child's future.

This book is for...

- Any student struggling with basic math
- Any *adult* struggling with basic math
- Parents with children ages 2 to 12
- Homeschool teachers
- Public school teachers

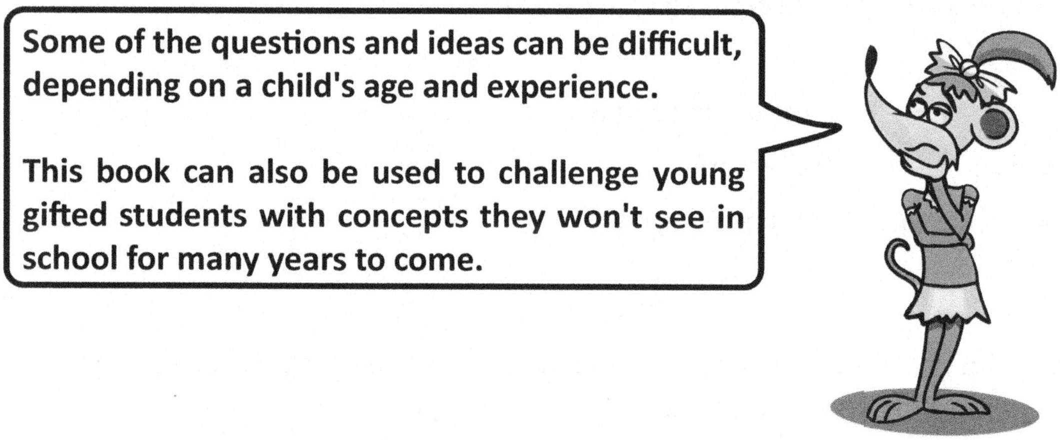

Section I:

Adult Learning

Before you start talking math with children, there are some important ideas to know...

1

Make it Concrete (Turn the Dial)

When your Internet stops working and you call the company for help, they always tell you to do one thing first: restart the equipment. That is a good first step when troubleshooting electronics!

In the same way, if your child has trouble understanding math, the first step should almost always be to **make the problem more concrete.**

Abstract	Concrete
Electricity	Light bulb 
Air	Inflating balloon 
Chemical reactions bonding sand & rock	Actual Concrete 
3	Three apples 

The last example might surprise you, but the idea of "three" is just as abstract as the idea of electricity. You can't see "three," and you can't hold "three" in your hand.

You CAN see the symbol that represents three (it's 3).
You CAN hold "3 apples."
You CAN "clap 3 times."

But, by itself, "three" is just an idea. It is abstract. You may be tempted to start working with numbers right away, but remember they are still brand new to kids. Adults have been using numbers for decades, so their meaning seems obvious to us.

Which Glass Holds More?

Here is a famous example of an idea that is obvious to adults but takes time for a child's brain to develop enough to understand.

I can fill a tall skinny glass with orange juice...

...and show that the same amount will fill a short, wide glass.

Children below a certain age will tell you that the TALL glass holds more juice, even though you just demonstrated that the glasses were equal.

You can TELL children that the glasses hold the same amounts—and they might even nod along—but they won't really believe you until they are old enough to understand that "taller" doesn't always mean "more."

Amounts Stay the Same

This idea is called "conservation" and means that the amount of something stays the same even if it is rearranged. You can count 5 cubes and hold them in your hand (concrete). What you CAN'T hold in your hand is the *idea* that the number of objects will stay the same no matter how they are moved around (abstract).

Complex ideas take time and experience to learn, so you can't just tell children about "conservation" and expect them to understand it. Give them more time to learn these "obvious" ideas.

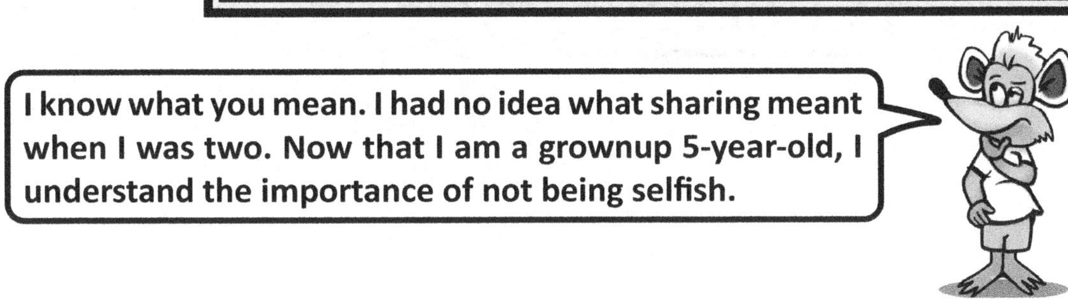

Moving from Abstract to Concrete

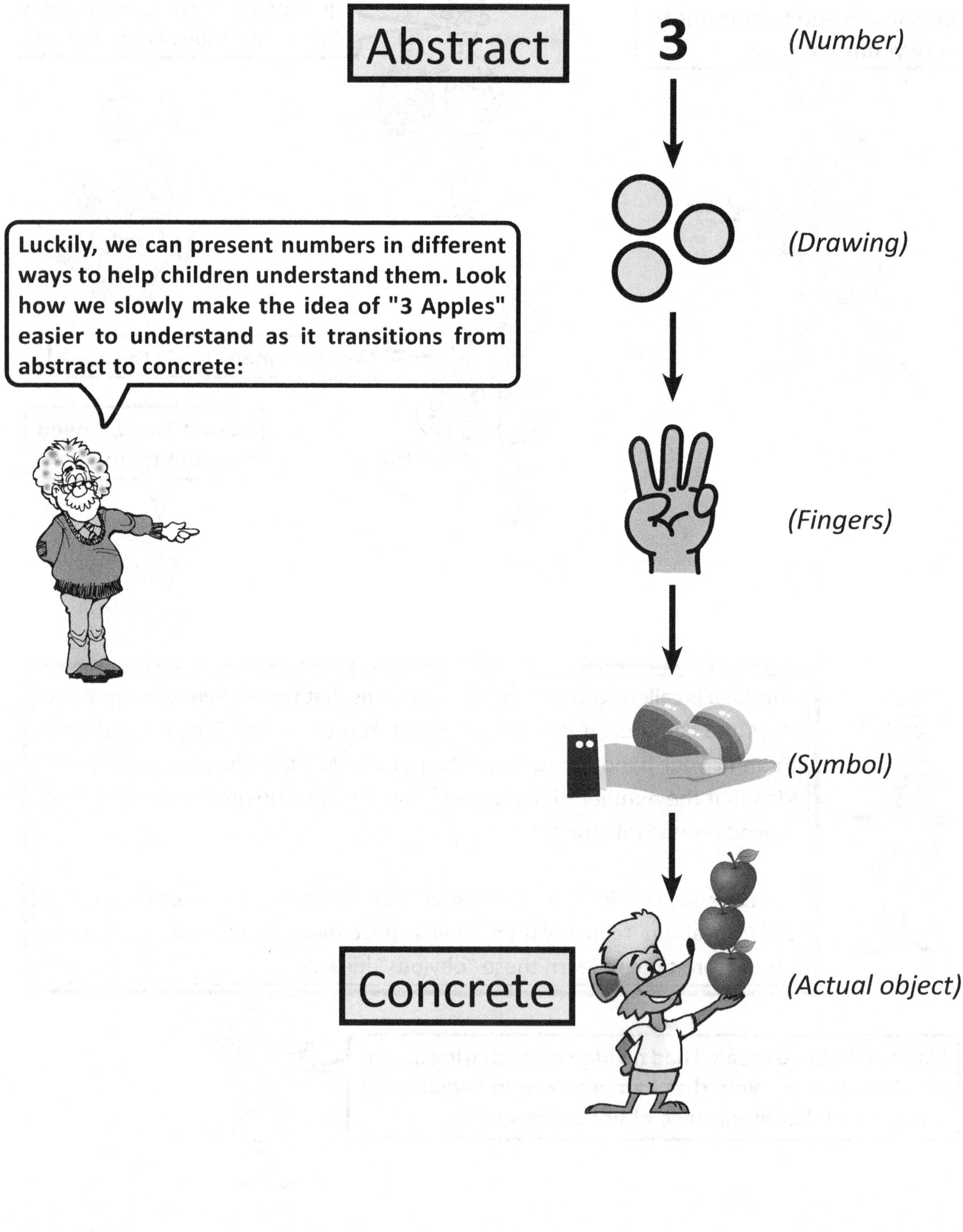

The Abstract Dial

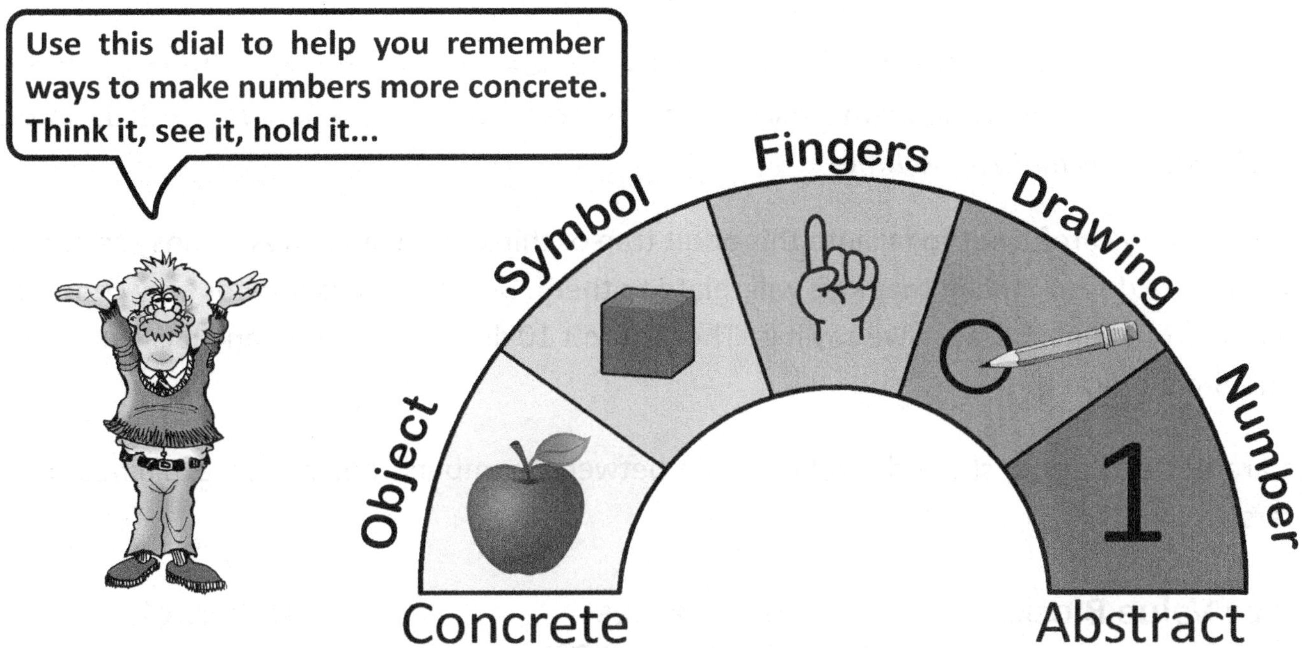

Here are examples of how a simple math problem can be represented in different ways using the Abstract Dial....

My shopping cart had 3 apples in it, then I grabbed 2 more. How many apples do I have now?

Actual Object

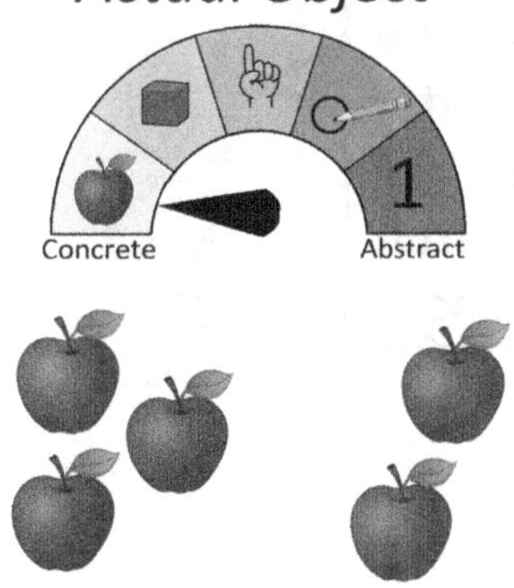

Symbol

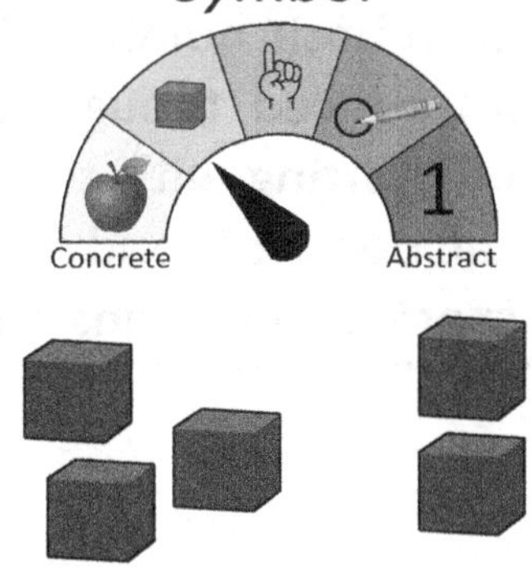

Fingers

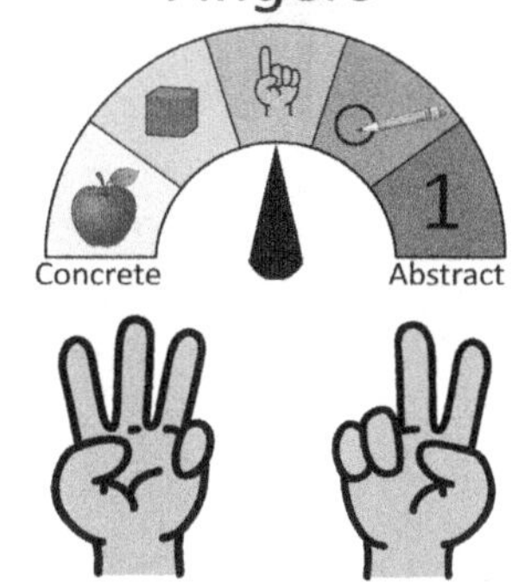

Drawing

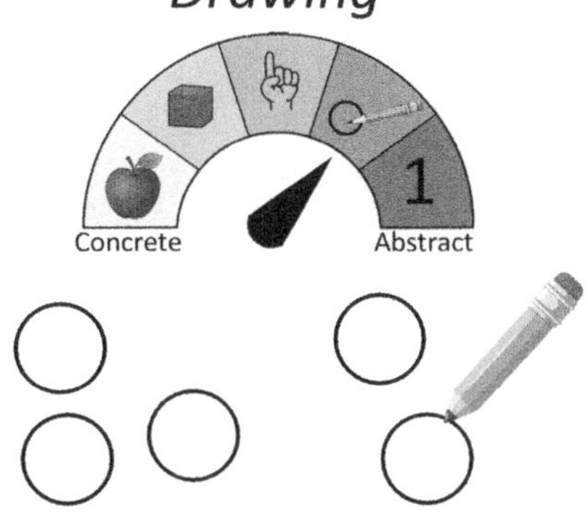

Numbers

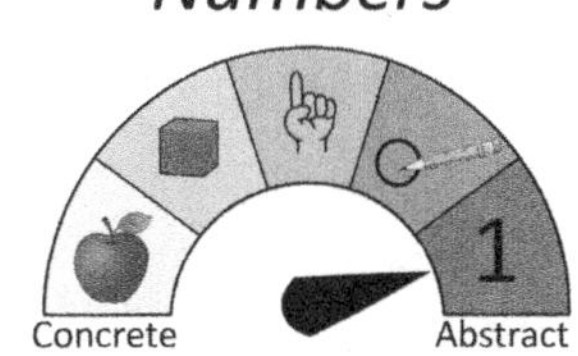

3 + 2 =

The Money Trap

Some parents think that an "obvious" way to make numbers more concrete is to relate them to money. They think, *"Everyone knows that a dime is worth 10 and a penny is worth 1, so using coins will teach kids math in no time!"*

We must be very careful not to fall into this adult trap of thinking. Just because coins are common in our *adult* life doesn't mean that kids will relate to them. Besides, coins are *not* very concrete! A dime is worth 10 cents because we say it is. There aren't 10 dots on the coin, and it is even smaller than the 5-cent nickel.

There are better ways to show the relationship between numbers that will be explained in later chapters:

Place Value Blocks

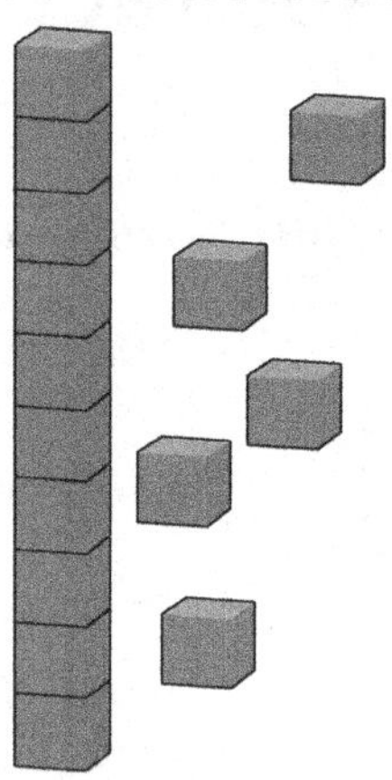

Ten Frames

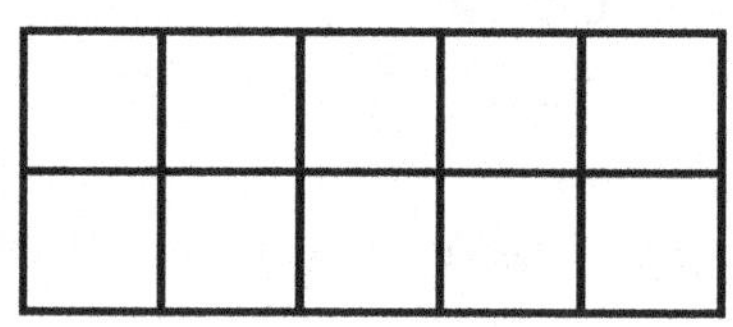

Rekenrek

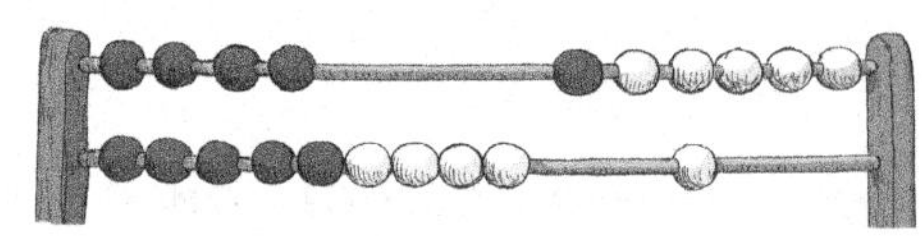

Even basic numbers (2, 5, etc.) should be introduced thoughtfully. You are better off starting with something that children can both *hold* and *relate to*. Crackers, fruit snacks, and raisins are all great options.

2

Why Rote Memorization Fails

Do you know the next line of this verse? If you don't speak Spanish, you probably won't remember that the answer is "Prospero año y felicidad." Songs in a foreign language are almost impossible to memorize, because to non-native speakers the words have no meaning.

P	G	P	I
L	C	I	P
A	O	N	E

P	I	G		C	O	I	N		A	P	P	L	E

As adults, it makes sense that children should easily memorize their addition and multiplication facts. After all, most of us know them, so how hard can it be? Apparently it can be very hard, considering how often children struggle to learn them. To children (and some adults) numbers are a *foreign language*. They see no meaning behind them, so they are left trying to memorize hundreds of different symbol combinations. This frustration leads people to think they are "stupid" or "just don't get math."

What adults see

$$7 + 3 = 10$$

What kids see

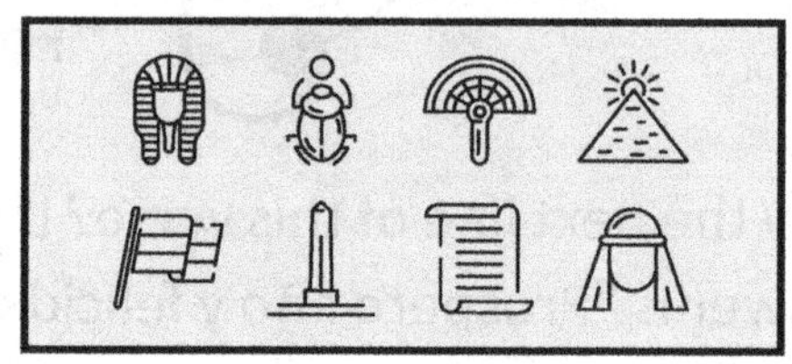

Prioritizing Procedures

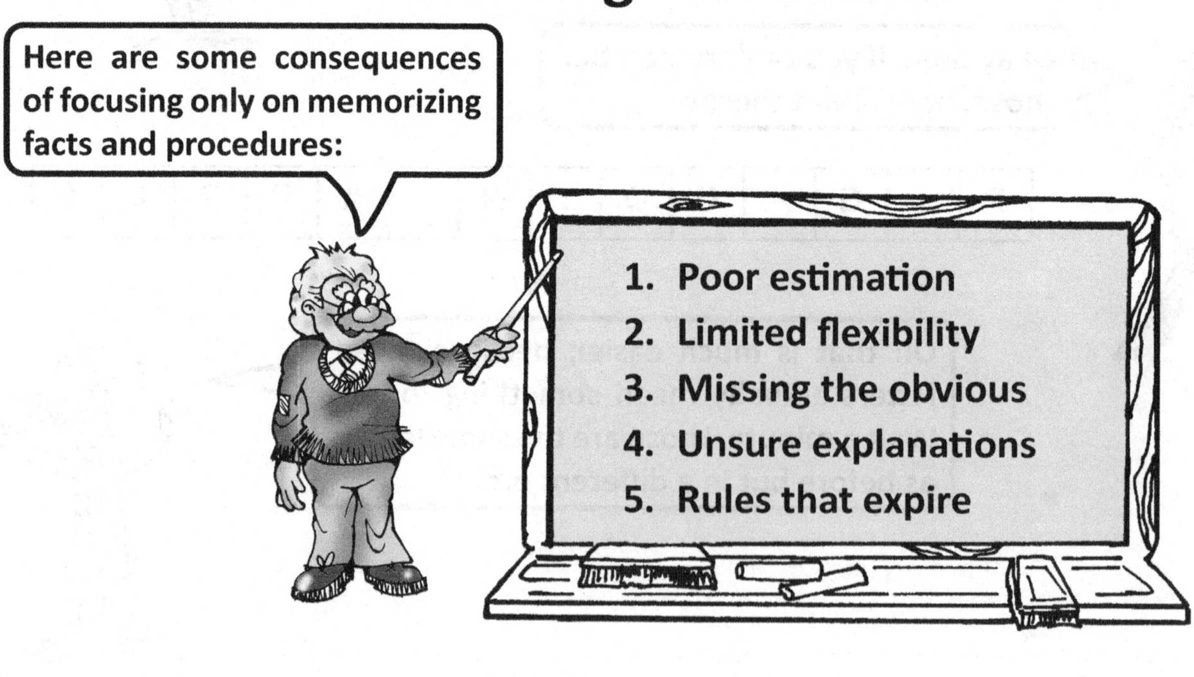

Poor Estimation

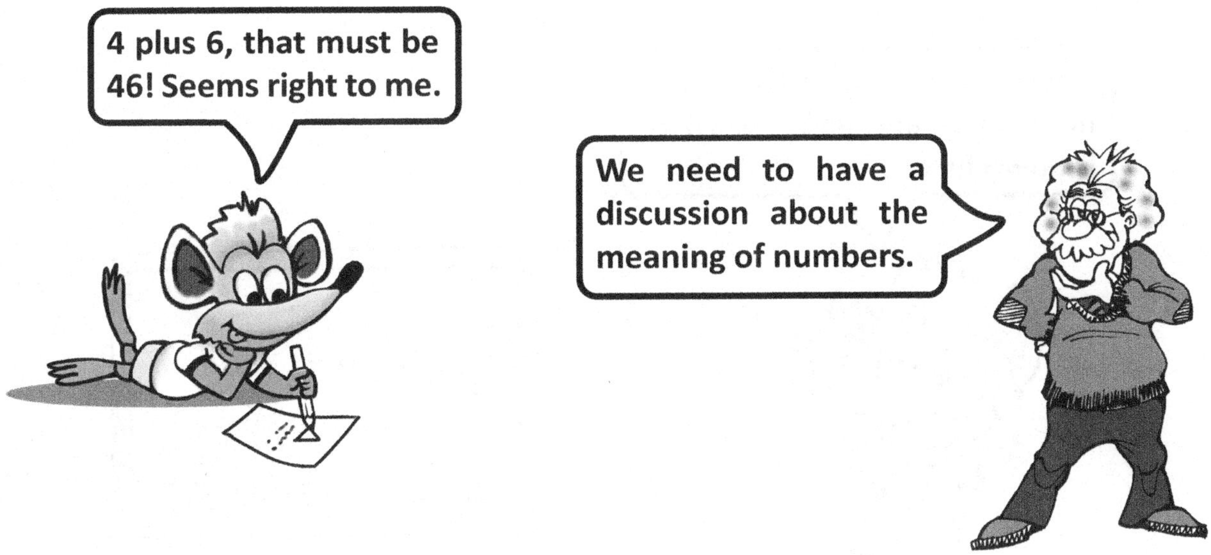

Limited Flexibility

Missing the Obvious

Unsure Explanations

Rules that Expire

Here is a rule that 1st graders often learn for word problems. This rule expires around the end of second grade. It doesn't really work after that.

"Just add all the numbers!"

Some kids are taught to "look for keywords" in math problems. This works less and less as students get older, and it limits their mathematical thinking. Do NOT teach keywords for story problems. Instead, talk about what the problem is asking or draw a picture.

Here are a couple more examples of instructional techniques that are taught in schools today that rely on math "tricks" which can lead to the correct answer but limit how students ***think*** about numbers.

"Add 10" means go up one row on the number chart

31	32	33	34	35	36	37	38	39	40
21	22	23	24	25	26	27	28	29	30
11	12	13	14	15	16	17	18	19	20
1	2	3	4	5	6	7	8	9	10

Fingers show multiplication facts for 9

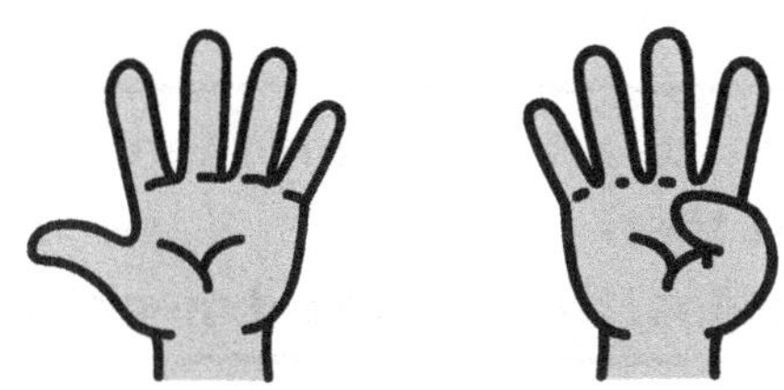

Memorizing facts and procedures works for *some* people, *some* of the time. But if children don't understand what numbers mean, they will struggle with math in school AND in life.

To learn more, read: 13 Rules That Expire by Karen Karp, Sarah Bush, and Barbara Dougherty (2014)

Road Blocks

Imagine you are driving through a brand new city and construction blocks the road. If you are not familiar with the area, you will have no way to think of different paths around the obstacle and it would be easy to panic! But what if this happened in your hometown? Almost instantly, your brain would start to picture new options for how to get to your destination.

Something similar happens when children get stuck on a math problem. Those who don't have a sense of what numbers mean won't see the connections that tie them together into a meaningful road map. For example, look at some of the different ways to see the number 18:

It is very important to teach children what numbers mean—even if you are a little unsure about numbers yourself. The next few chapters will give you an idea of how to approach your teaching, whether it is in a classroom or at the kitchen table. This will help you set the stage for all the learning that will happen over the next several years.

3

The Right Environment

Everyone learns in different ways, but we have to be especially careful with how we talk about math to young children if we want to keep them engaged. Some teachers give long lectures and then expect students to do worksheets by themselves, but learning math this way doesn't always work. Some math ideas come to kids naturally, but others never fully develop without help.

Remember these points to help keep children on the right path:

- Keep lessons short and active
- Use movement and sound
- Have conversations
- Embrace mistakes
- Teach *ideas* before vocabulary

Keep Lessons Short and Active

3-year-olds CAN talk about math, but not for very long. Attention can drift after 5 minutes or after 30 seconds—it all depends on the child.

Use Movement and Sound

"Three" can be demonstrated with 3 drum hits, 3 jumps, or 3 claps. Not only does this keep the activity interesting, but it also expands the idea of what "three" means.

Have Conversations

This applies more to children who have basic language skills (usually at least 3 years old). Go beyond the basic "that's right" or "that's wrong" and encourage conversations about *why* a child answered a certain way. Remember you can also ask follow-up questions for right answers (otherwise your child will learn that a question from you means he's wrong).

"How do you know?"

"Try that again."

"Show me another way."

"What does ___ stand for?"

"How did you get that answer?"

"How else could you find the answer?"

"Count again to double check."

"Draw me a picture."

"Show me with these objects..."

Embrace Mistakes

Correct Answer

What does this correct answer tell us about Jill's math ability?
Only ONE of these is probably true:

1) Jill knows all her facts and can explain why.
2) Jill has memorized a few facts and this is one of them.
3) Jill used her fingers and toes to count each number one by one.
4) Jill solved this mentally, but it took her a long time.
5) Jill copied off her partner.

Incorrect Answer

Now, what does this INCORRECT answer tell us about Sam's math ability? **ALL** of these are probably true:

1. Sam solves problems left to right.
2. Sam works by rote (step-by-step) without thinking about what he's doing.
3. Sam does not understand basic place value concepts.
4. Sam does not double-check his answers to see if they make sense.

Children also need to understand that challenge and frustration are a normal part of the learning process (and life). This means they shouldn't be stuck in a pattern where "everything is easy."

Teach *Ideas* before Vocabulary

It is tempting to teach a list of facts and move on, but that doesn't help children understand what they mean. Making sure they have a solid foundation will save you a lot of re-teaching time later on. Give young children *basic concepts*, not fancy vocabulary. The idea of "more" is much easier to understand than words like "addition." Math vocabulary **is** important, but it should not be the first step in learning.

Turn the Dial

If the child continues to struggle, see if you can turn back the Abstract Dial to make the problem more concrete (and easier to understand).

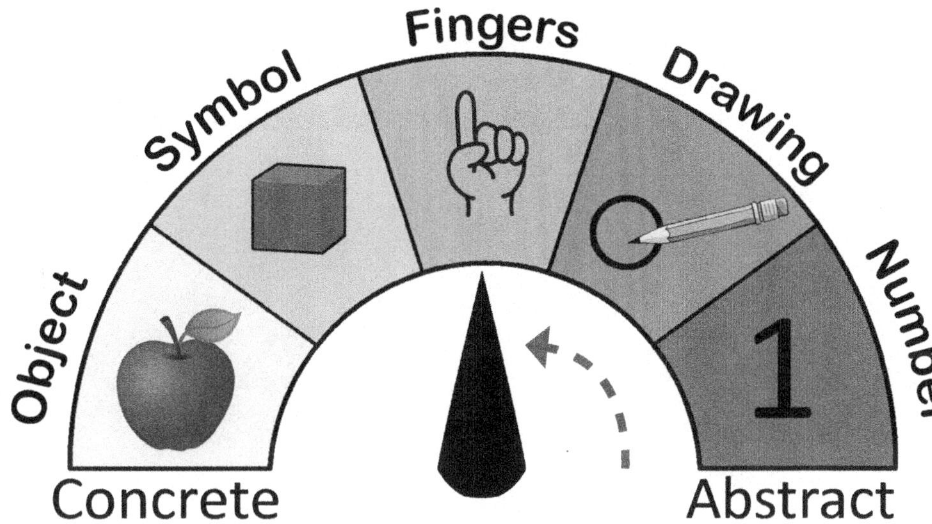

On the other hand, if a question is too easy for your child, you can make it more abstract by turning the dial to the right! The end goal should always be to work with written or spoken numbers.

Some Final Advice...

- The student should be doing most of the thinking, not you! Learning math is an active process, so be patient and try not to take over!
- Remove these sentences from your vocabulary: ***"I'm not good at math"*** and ***"I'm not a math person."*** Everyone is a person, and everyone does math. Set a good example!

4

Building Routines

The bad news about teaching math to two-year-olds is that they are not going to sit down with you for a 30-minute lesson about what a number means. The good news is that you have plenty of time to start!

The best way to introduce young children to basic math is to build routines throughout their day that give them the context for numbers and math ideas. This starts with you! The more you (as an adult) see math in your life, the more you can talk about it with children.

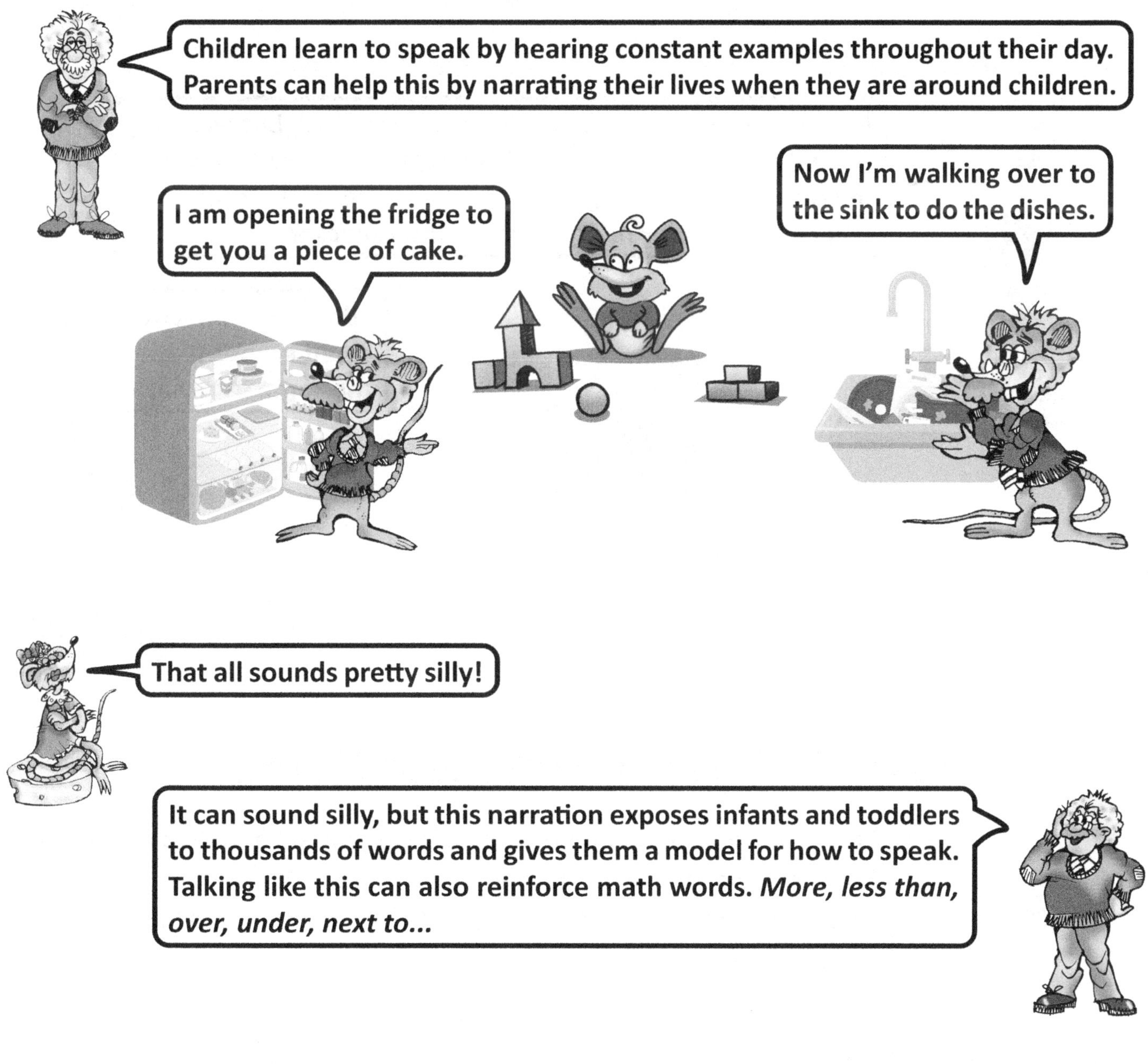

Applying this idea to math means that you will need to keep an eye out for small numbers and amounts during your day. When speaking to very young children, try to keep this limited to 1, 2, and 3 (these amounts are easier to see and understand). We are only focusing on *amounts* at this point, **not** numerals (*12, 14 ,7, etc.*).

*"Here are **TWO** crackers"*

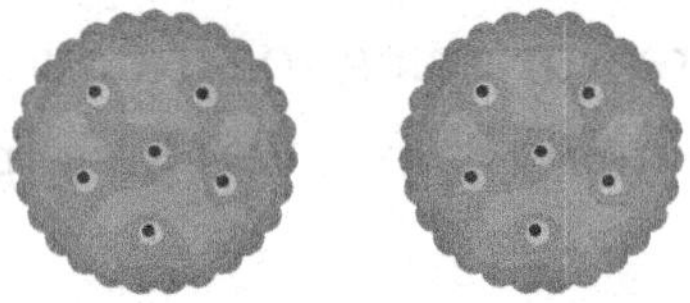

*"Bring me **TWO** shoes. ...yes, that's **ONE** shoe but I need **TWO** shoes."*

*"Do you want **ONE** vitamin or **TWO** vitamins?"*

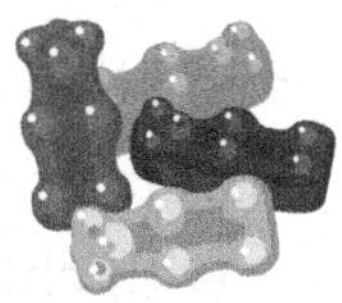

Here is another tip...use food! Not as a reward, but as a way to make numbers more real. People are more interested in learning about things they can relate to in their life, and what can a toddler relate to more than snacks?

I don't want to count, but I will count grapes to make sure I get the same as my little brother!

~~One~~ Zero is the Loneliest Number

Zero is an important number, but it is also abstract and hard to understand. We can see **one cracker**, but we cannot see **zero crackers**. Children must understand what 1 or 2 crackers look like before they can understand what "zero crackers" look like.

Universal Routines

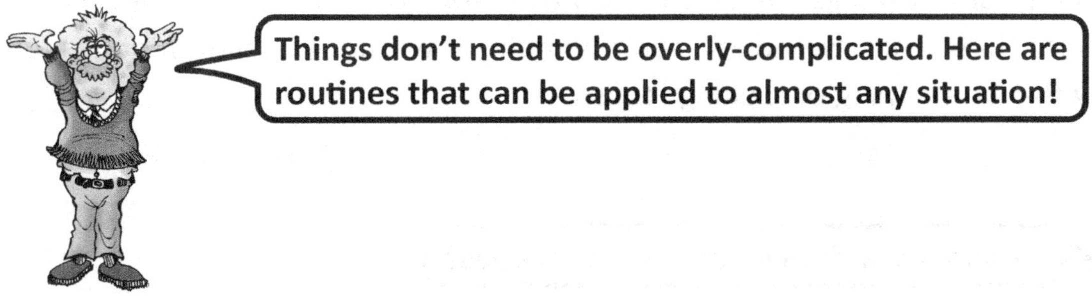

"What do you *NOTICE*?"

Everyone can "notice" something, so this question gets children talking about math ideas no matter how old they are. They will learn that there isn't a "right" answer, which helps them become more comfortable during math conversations. It helps to model things that *you* notice until your kids get the hang of it.

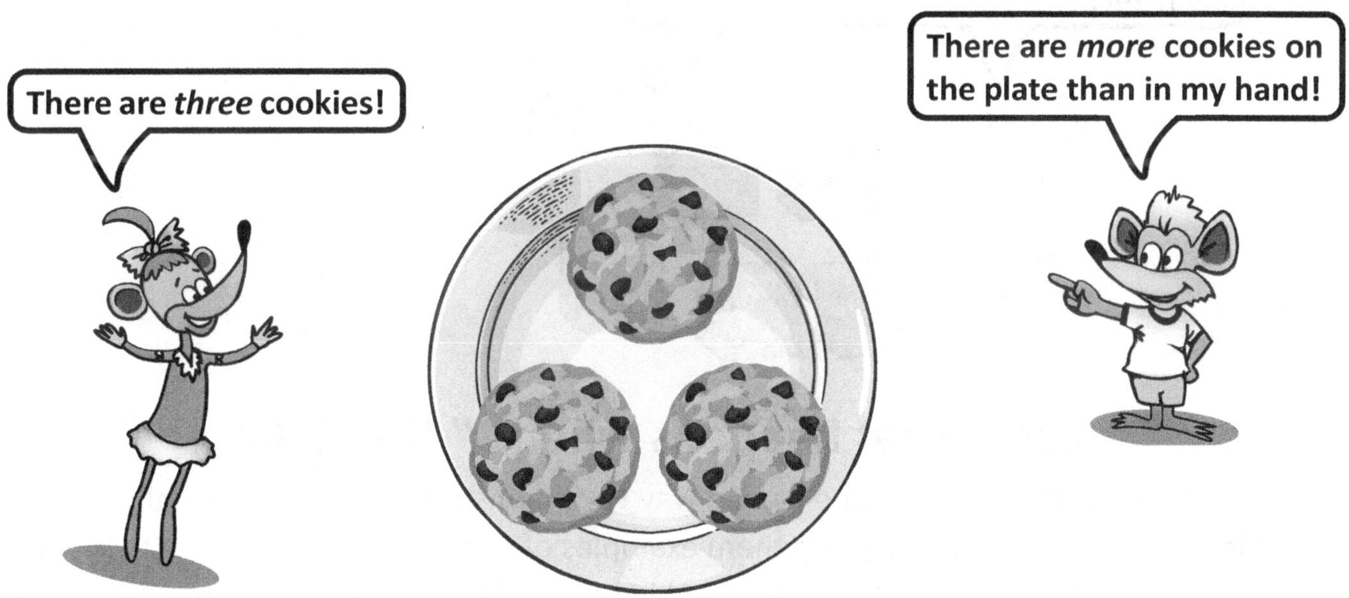

"What do you *WONDER*?"

Everyone has questions about things, which can also help start conversations about math. Just like when we "notice" things, our "wonders" can't be right or wrong, so children don't need to worry about saying the wrong thing or looking foolish.

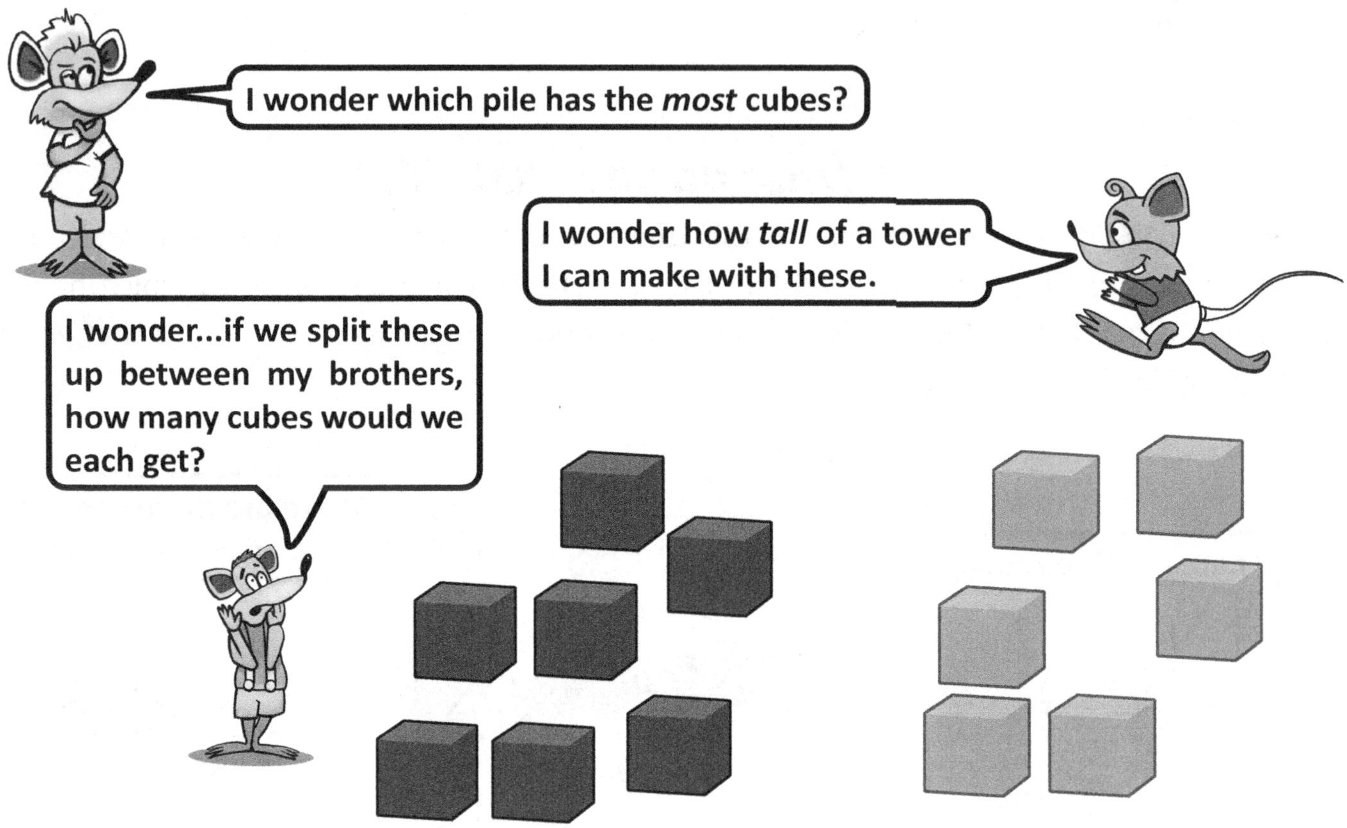

"Look what happens when..." (Modeling)

Pairing an *action* with an *idea* can be a good way to demonstrate something visually to children. This helps with conversations by giving them examples of math words and ideas without needing them to contribute each time. Modeling provides a nice break to the constant questioning we sometimes get in the habit of doing.

"Which one doesn't belong?"

Like the other routines, these conversations have no "right" answer. Here is a non-math example:

"Which one doesn't belong?"

This one doesn't belong because...it is the only farm animal

This one doesn't belong because...it is the only animal that is jumping

This one doesn't belong because...it is the only animal laying on it's back

This one doesn't belong because...it is the only animal wearing a collar

I like how I can think of different reasons for each square. There are a lot of options, which makes the conversation fun!

Of course, we can do this with math ideas, too!

"Which one doesn't belong?"

This one—the dots aren't in a shape

This one—the dots are lighter

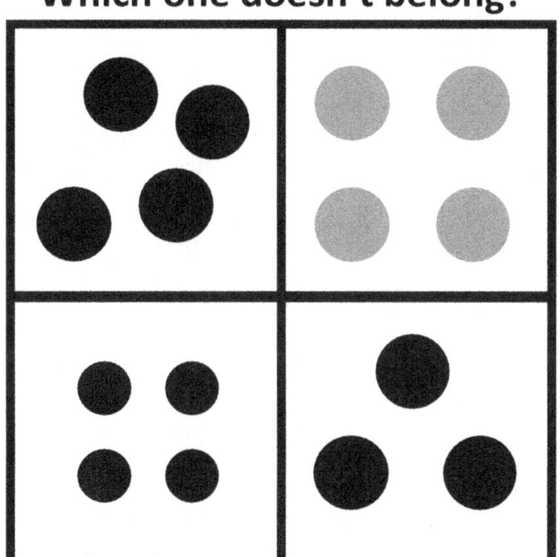

This one—the dots are smaller

This one—there are only 3 dots

Illustrative Math is a new curriculum that uses these routines and others. Learn more at: https://illustrativemathematics.org/

5

Problem Overview

The questions in this chapter are not intended to be solved now. They are an illustration of the concepts that children will learn as they work through the rest of the book.

Number Sense

1) **(Sorting and More/Less)**
Are there more **pieces of cheese** or **rocks**?

2) **(Recognizing Small Amounts)**
Show these apples for 2 seconds, then cover them with your hand.
How many apples were there?

3) **(Counting Out loud)**
Count to ten.

4) **(Counting Objects)**
Count the dogs. Tap them with your finger.

5) **(Counting to find "How Many")**
Count the squares from left to right. How many will there be if you count them the other way?

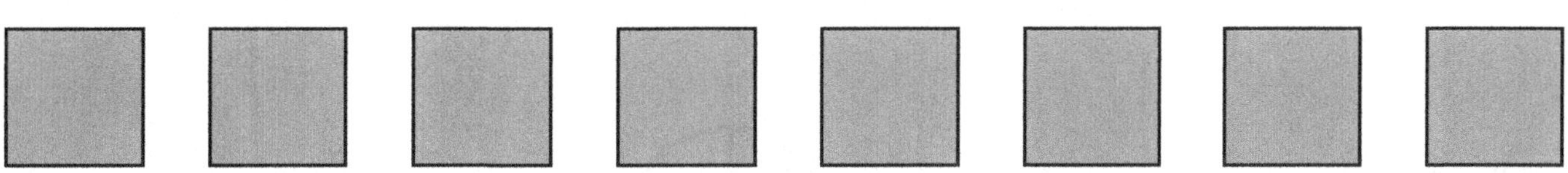

6) **(Joining and Separating Numbers)**
Draw eight shapes, some circles and some squares. What is another way you could draw them?

Applying Number Sense

7) **(Addition and Subtraction)**
A full deck of cards has 52 cards in it. Kevin's deck has only 42 cards. How many cards are missing?

8) **(Place Value)**
In expanded form, the number 35,625 is equal to: 30,000 + 5000 + 600 + 20 + 5
What is the number 88,888 written in expanded form?

9) **(Multiplication and Division)**
Eggs usually come in cartons of 12 (called a dozen). If you need to carry 48 eggs, how many cartons do you need?

10) **(Math Facts: Addition and Subtraction)**
"Number Bonds" are a way to show fact families, or the relationship between adding and subtracting numbers. Complete the rest of these Number Bonds:

11) **(Math Facts: Multiplication and Division)**
Explain how you can use the fact 6 x 10 = 60 to tell you what 6 x 5 is.

12) **(Fractions)**
What fraction of the circle is shaded?

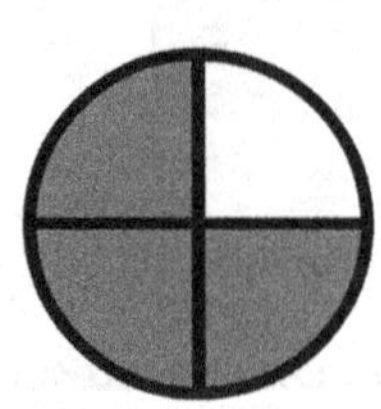

13) **(Percents)**
What is 2% of one dollar?

14) **(Decimals)**
Put these in order from smallest to largest: **$\frac{3}{10}$** **.28** **.29**

15) **(Estimation)**
If $98 is to be divided among 4 people, about how much money will each person receive?

Advanced Concepts

16) **(Language of Algebra)**
Henry weighs 5 pounds more than twice his dog's weight. If Henry's dog weighs *n* pounds, how much does Henry weigh (in the language of algebra)?

17) **(Balancing Equations)**
Find n: $6n + 100 = 103$

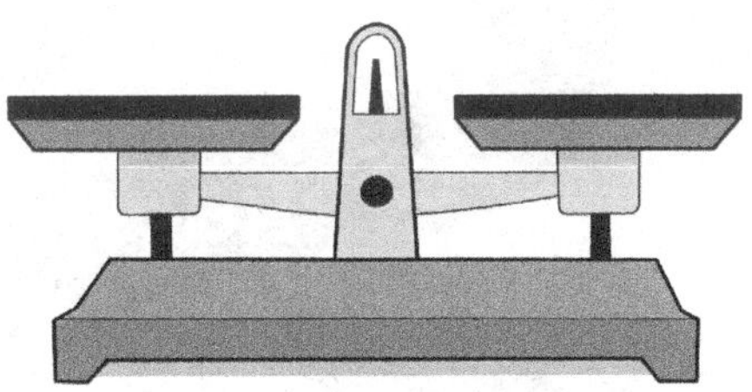

18) **(Ratios)**
8 people have enough food for a 6 day hike. If 8 more people join them, how many days will the food last?

19) **(Speed, Distance, and Time)**
If a whale swims 120 miles in 10 hours, what speed was it traveling at? (miles per hour)

20) **(Negative Numbers)**
If the temperature was 25 degrees in the afternoon and then dropped 26 degrees by the time it was midnight, what was the temperature at midnight?

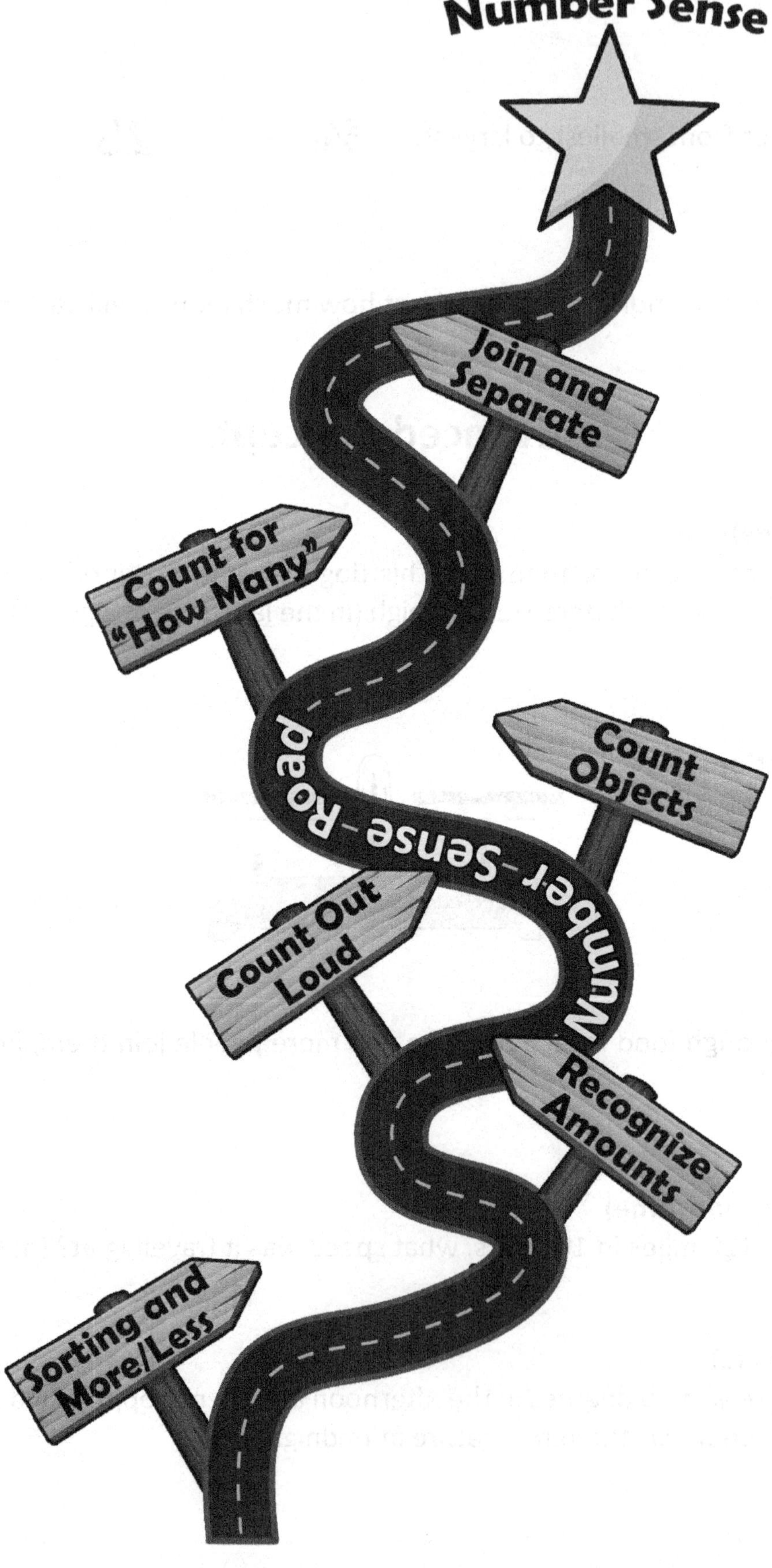

Number Sense
Join and Separate
Count for "How Many"
Count Objects
Number-Sense-Road
Count Out Loud
Recognize Amounts
Sorting and More/Less

Section II:

Number Sense Road

Understanding what numbers mean lets us use them flexibly and discover patterns between them...

6

What is Number Sense?

Experienced cooks are *flexible* in the kitchen and instantly know which ingredients can be substituted and which ones should not be. They can read a recipe and quickly think of alternatives for any missing ingredients.

People without this skill get stuck if they can't follow the "rules on the box" exactly. They don't see an alternate route and miss obvious solutions like *just use the skim milk instead.*

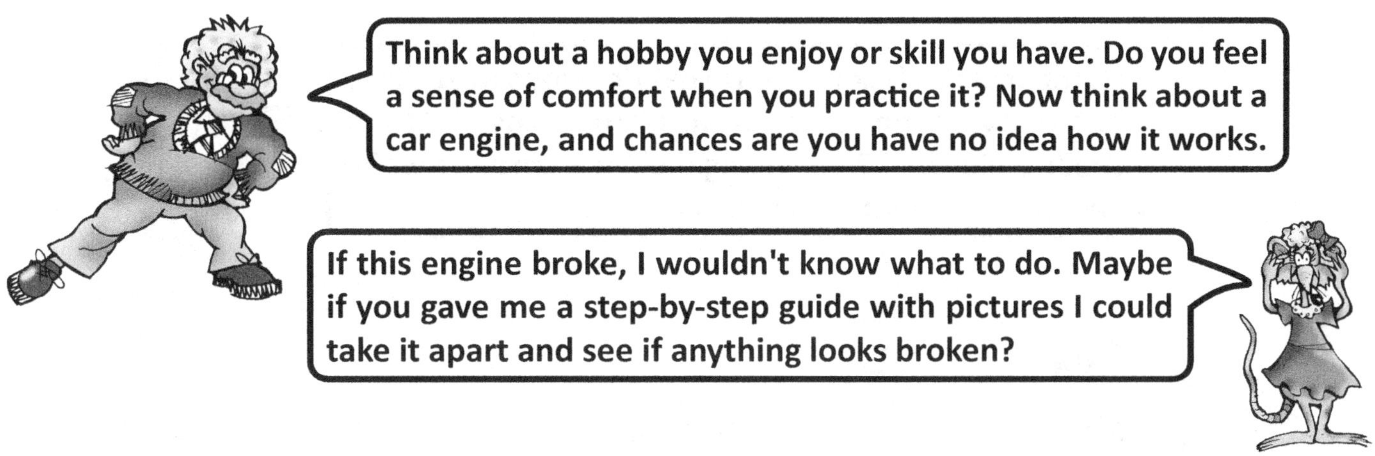

That might work for a while, but one unclear picture could stop the entire process. Most mechanics thought engines were complicated at first, but eventually they developed good *engine sense* and stopped needing perfect step-by-step guides to repair them.

Here's the good news: Numbers are much easier to work with than engines. To have Number Sense means to understand what numbers are and how they relate to each other.

5

5...that's how many fingers I have on one hand, and it's how many days my dad works each week. Five crackers is 1 more than 4 crackers. If I double 5 I get 10......I think I am really starting to understand numbers!

Some people say they don't understand math at all. Were they born without number sense?

Number sense is something that develops over time, but everyone has some of it, even young children. Don't believe me? Try giving a child 2 crackers while you give his brother 6 crackers.

Numbers Mean Different Things

My mom says it is TWO o'clock, but we only have ONE clock. That doesn't make any sense.

Now you see why numbers can be so hard to understand. When you see a "2" it can mean *two of something, second place*, or it can just be a label that means nothing special at all.

Amount

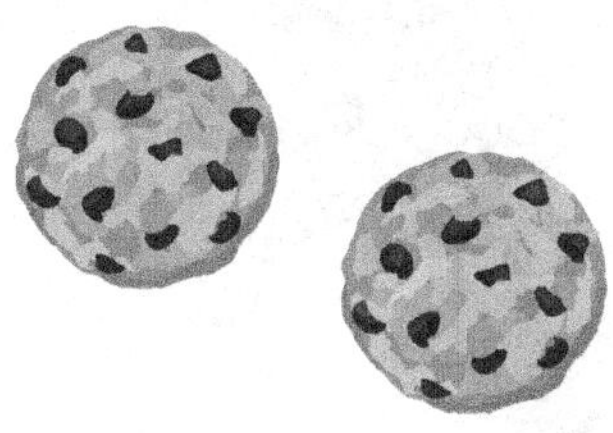

Order

2nd place!

Label

Numbers are *symbols*. Look how many different meanings a 2 can have!

Room 2

Jump twice

2 minutes

Two more bites of dinner before dessert

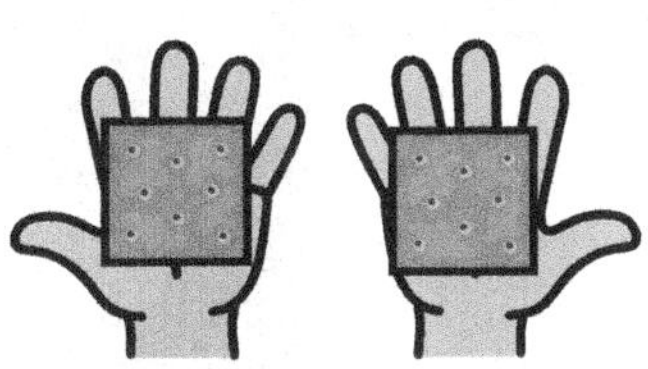

2nd place

2 degrees hotter than yesterday

"Amounts" are easier to understand than other types of numbers. This especially applies for things we can see or hold.

"2 apples" is easier to understand than "2 degrees Fahrenheit."

The Road to Number Sense

Here are the skills that everyone needs on their road to building Number Sense (the next few chapters will explain these ideas in detail).

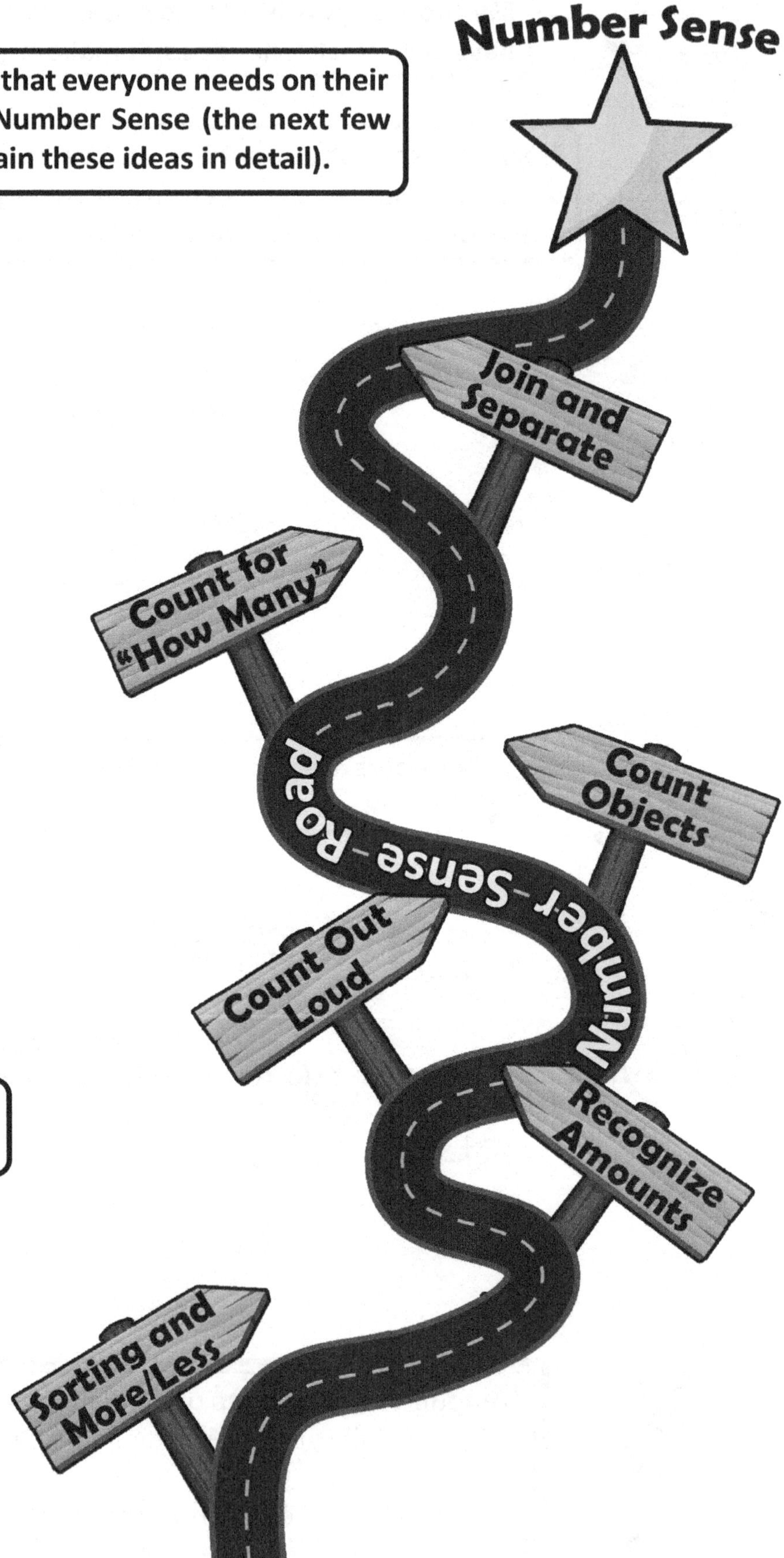

I better buckle up, this looks like a fun ride!

Sorting Objects

Recognizing More and Less

Recognizing Small Amounts

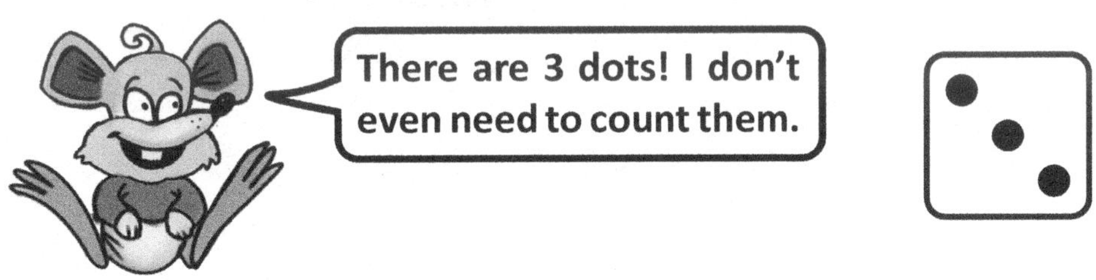

Verbal Counting

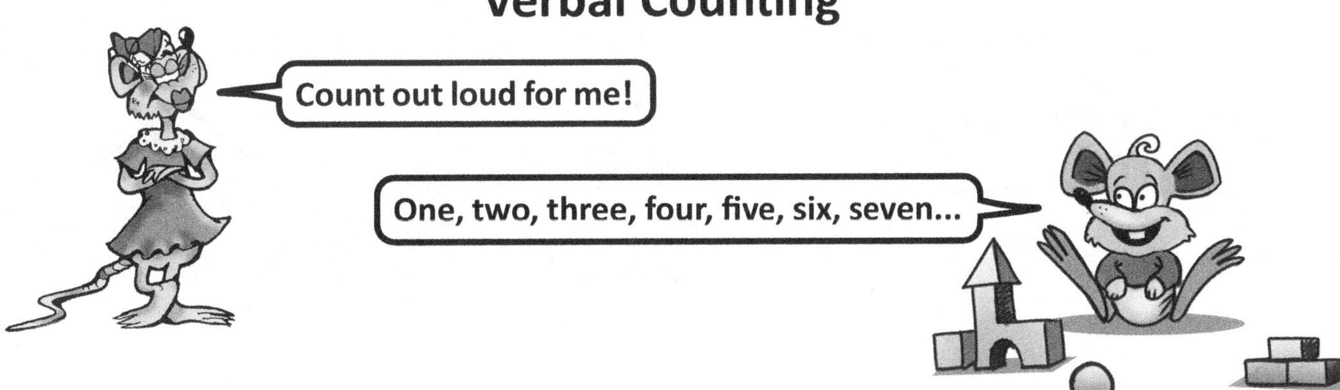

Counting Objects

Counting to Find "How Many"

Breaking Apart and Joining Numbers

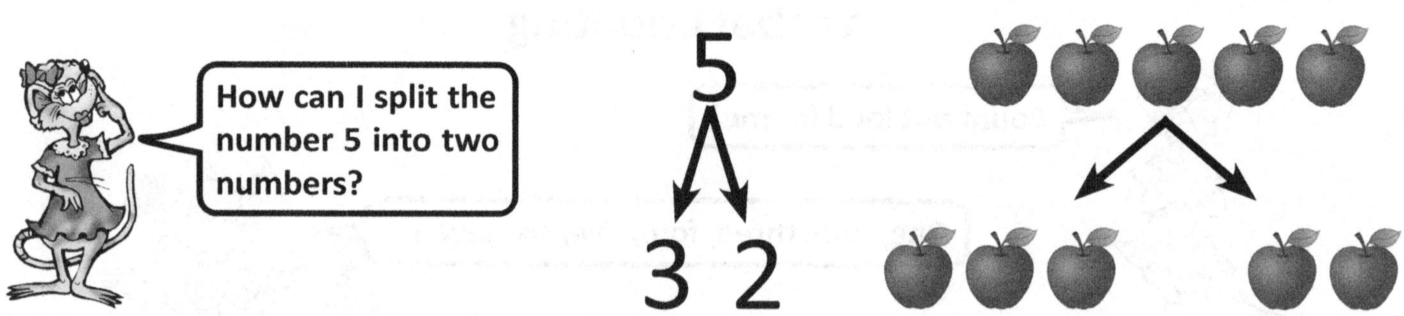

7

Sorting and More/Less

I thought we were learning about numbers! What does sorting objects have to do with math?

Math is about making connections, like connecting amounts to numbers. All of us grow up learning how to group different types of things.

I don't know what makes animals different. They just ARE.

Children are usually able to sort objects into categories before they can explain *why* they belong in different groups. Just like with animals, the more experience we have with numbers, the better we can group them:

odd — even   big — small   2 — 3

My toys — Your toys

Red blocks — Blue blocks

Loud voice — Quiet voice

Numbers — Letters

Big — Small

Squares — Circles

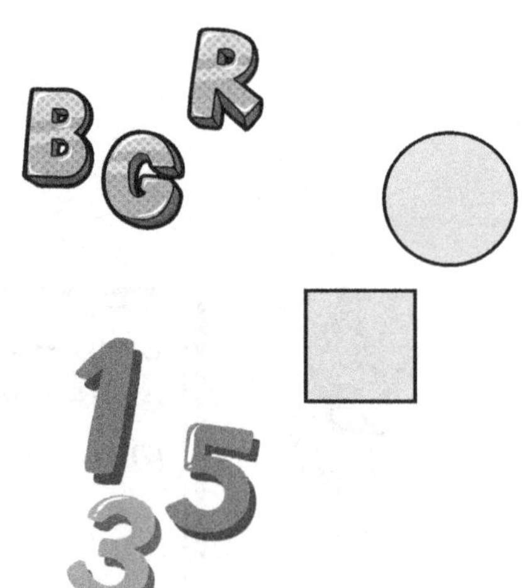

3 vs. 5

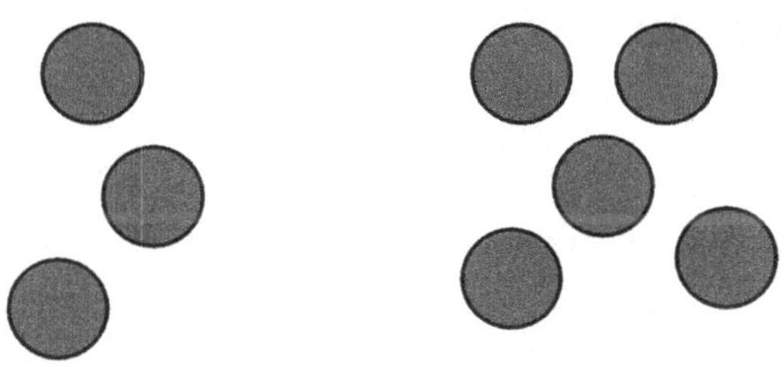

Less vs. More

When children practice sorting, they learn the differences and similarities between the objects.

Dogs and cats both have fur, but they are different in other ways. "2" and "5" are both numbers, but they are different in other ways. Children must *experience* the difference between numbers before they can truly understand them.

Sorting and More/Less

Level 1

When you try the following problems with children, keep in mind that younger children may respond better to real objects instead of pictures. Consider adapting these questions by using snacks, toys, or other objects.

1) Which of the cats below looks the same as this cat? 

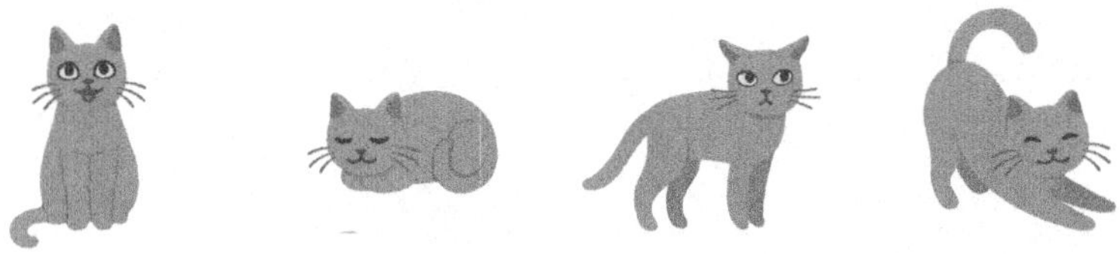

2) Which of the shapes below is the same as this shape? 

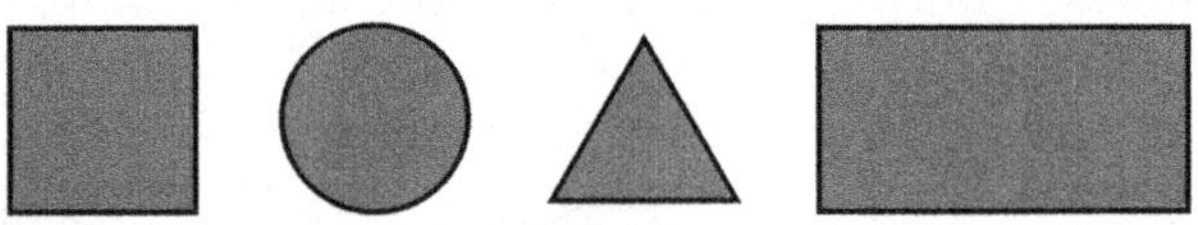

3) Which of the dogs below looks the same as this dog? 

4) Touch all the **big** squares below

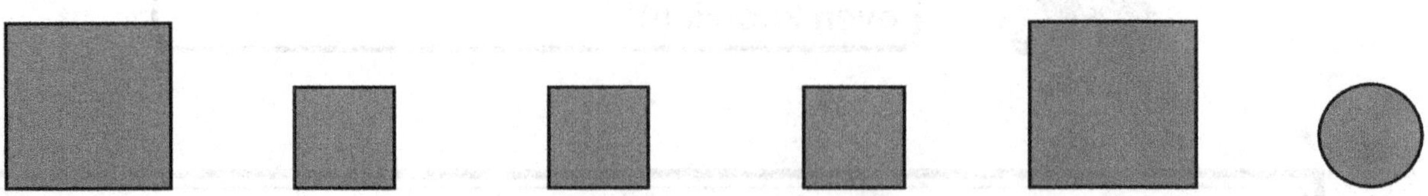

5) Which of the letters below is the same as this letter? A

B T A G

6) Which lions are moving?

7) Point to the rats that have something in their pocket.

8) Which number below is the same as this number? 8

3 5 8 4

9) Are there more **ants** or **circles?**

10) Are there more **pieces of cheese** or **rocks**?

Sorting and More/Less

Level 2

1) Point to the shapes that are big circles:

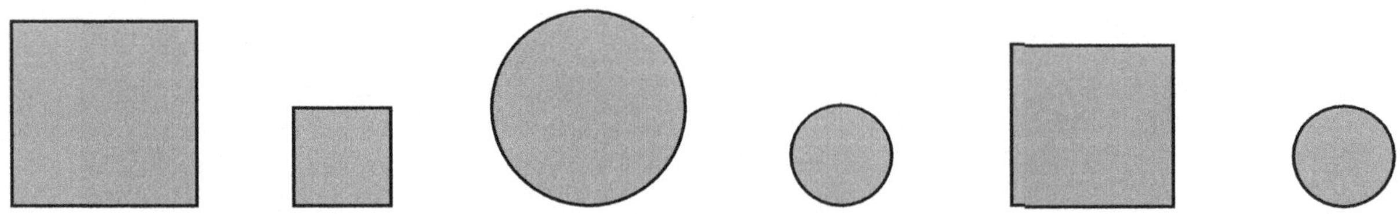

2) Which of the dogs are big **and also** sitting?

3) Which of the cats are small **and also** jumping?

4) Point to the letters that have another letter next to them.

AB H GH CD

5) Which rat is upside down **and also** pointing?

6) Point to the big light circle.

7) Which circle has more symbols in it?

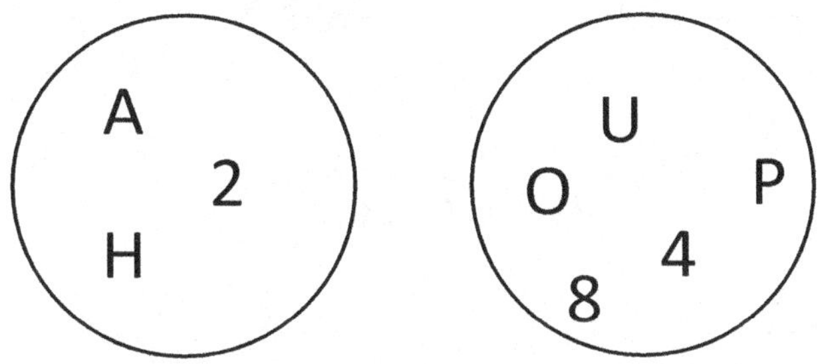

8) Which pile has more squares?

9) Which circle has fewer numbers?

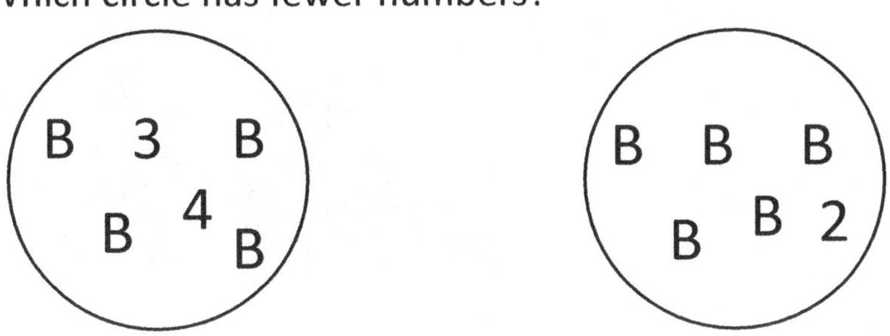

10) Which numbers are big **and also** black?

H 2 2 T G 7

Sorting and More/Less

Level 3

1) Point to all the circles that have a square inside them

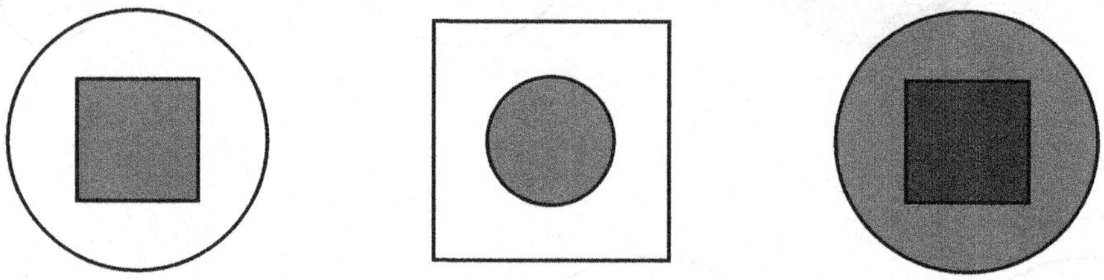

2) Which circle has more dogs in it?

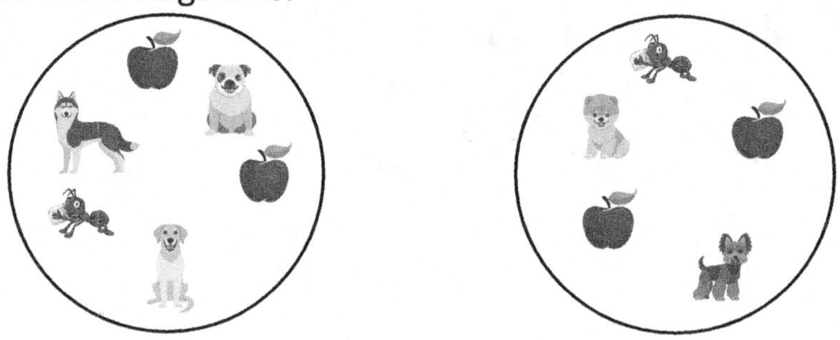

3) Which pile has the **least** number of cookies? Which pile has the **most** cookies?

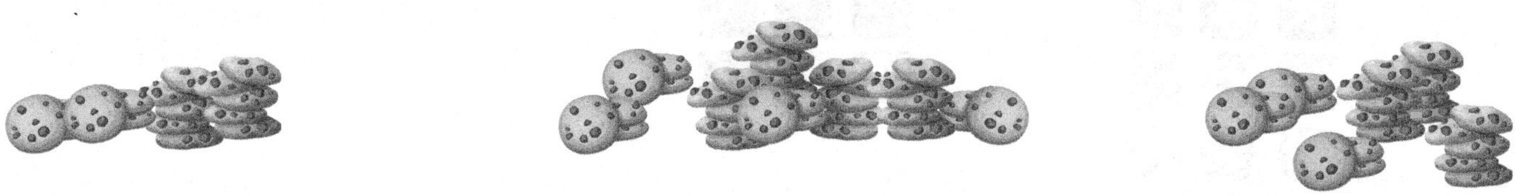

4) Which gorilla has more blocks to play with, the adult gorilla or the baby gorilla?

5) Which plate has the most apples (or are they the same)?

Sorting and More/Less

Genius Level

1) Draw lines to put the mittens and gloves in the correct boxes.

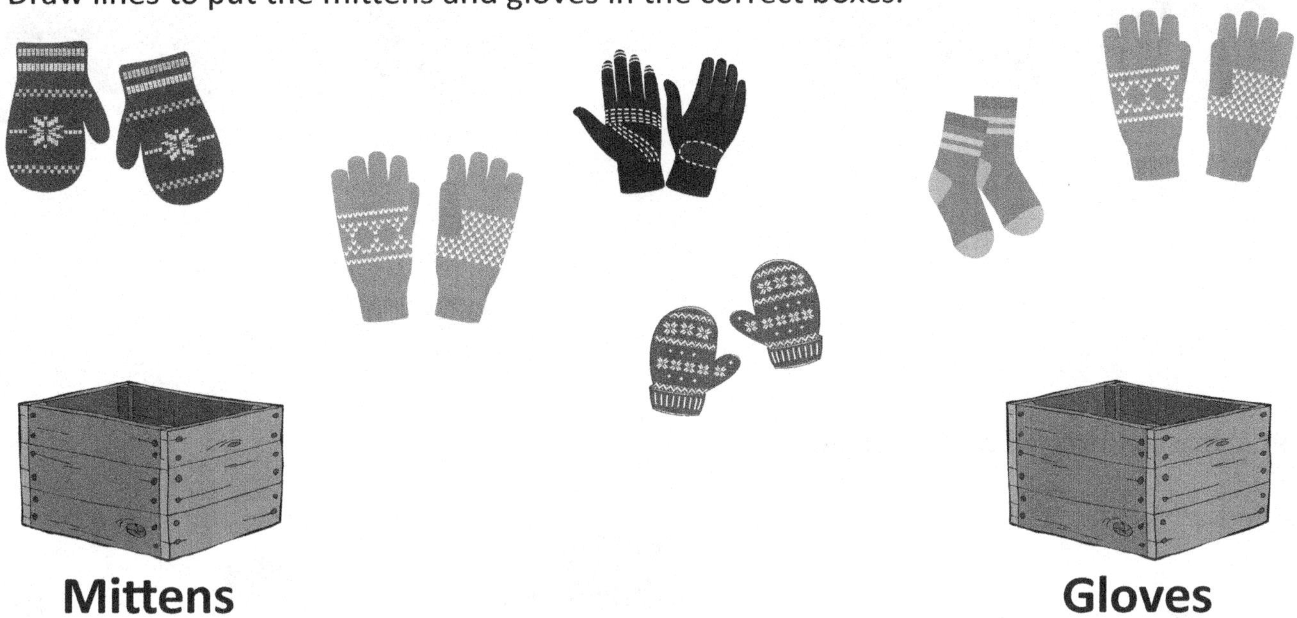

2) Point to the **smallest black circle**

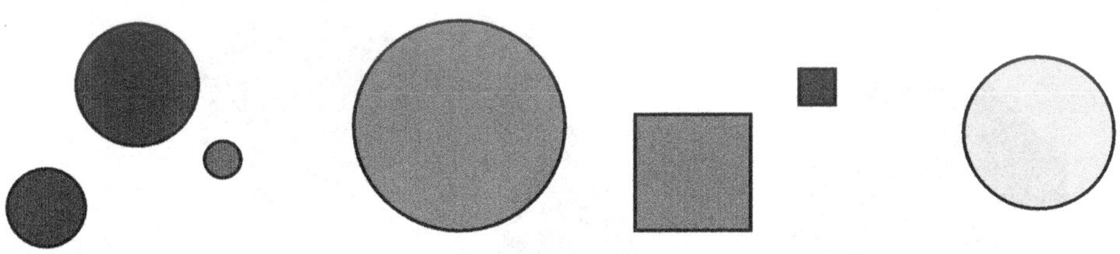

3) What group has more legs: 3 spiders or 10 people?

4) Which side has more apples? (Some apples are hiding under each bowl.)

5) Point to the biggest black hexagon.

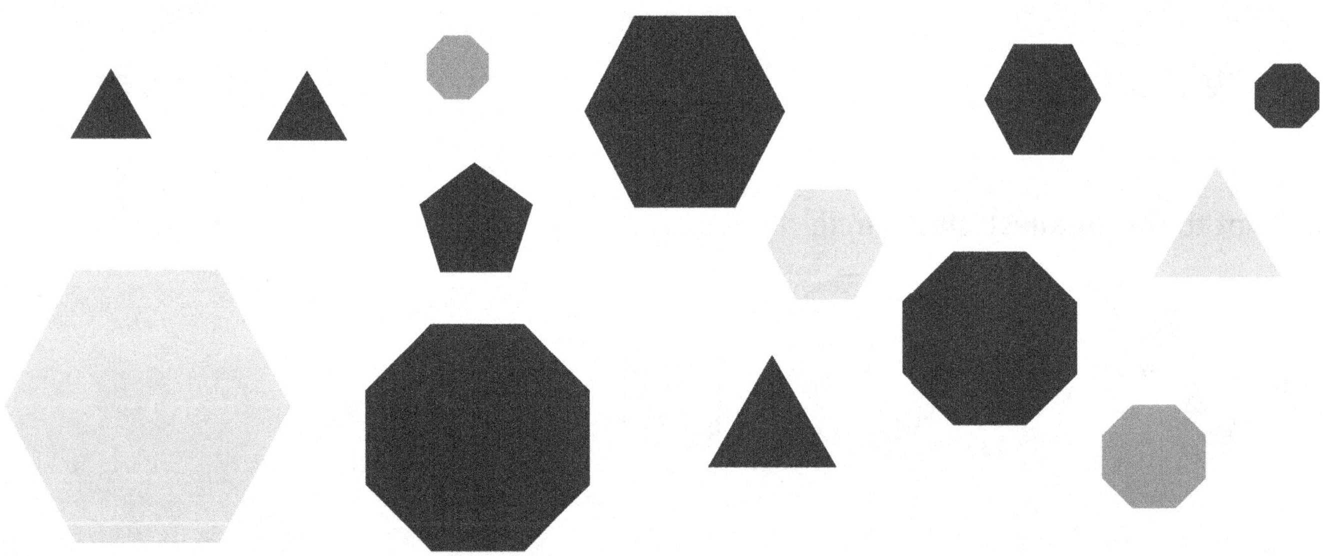

8

Recognizing Small Amounts

When I ask my son to count small amounts—like 3 or 4 raisins—he tells me the amount right away without needing to count them. Is he a superhero?

Recognizing small amounts is something all of us do as we get older. This is called "subitizing" and it's one of the early steps for number sense! Keep in mind it is limited to SMALL amounts: starting with up to 3 and eventually up to 5.

(It's pronounced like "soobitizing," but you don't need to remember the word, just the idea.)

I can roll a die and know how many dots there are without counting them. Maybe I'm a math superhero too!

Sure, you *could* count small sets of objects one-by-one, but why waste the time if our brain can recognize the amount instantly? You can test this yourself by rolling a die and seeing if you can recognize the amount without counting the dots.

Getting to Automatic

Imagine how long dice games would take if we had to count all the dots after rolling them!

This ability to "no count" small amounts develops naturally, but sometimes kids need to be pushed to say the amount without counting. Try these tricks to build the "no count" skill (remember this is only expected for up to 3 or 5 objects depending on the child's age).

"Don't Count"

How many apples are here? Tell me *without* counting them.

Cover It

Show the objects for a couple seconds, then cover them and ask how many there were.

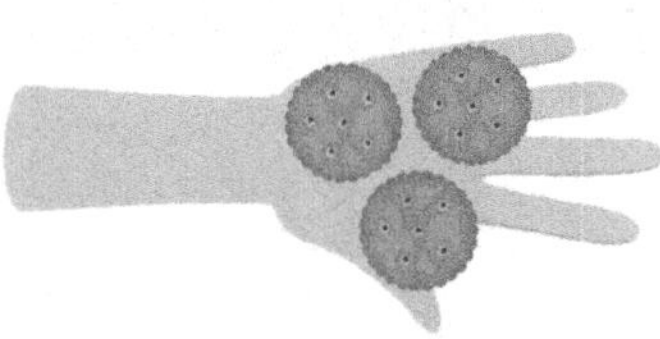

(1-2 seconds)

"How many crackers do I have?"

"Which is Different?"

Which group of dots is different from the others?

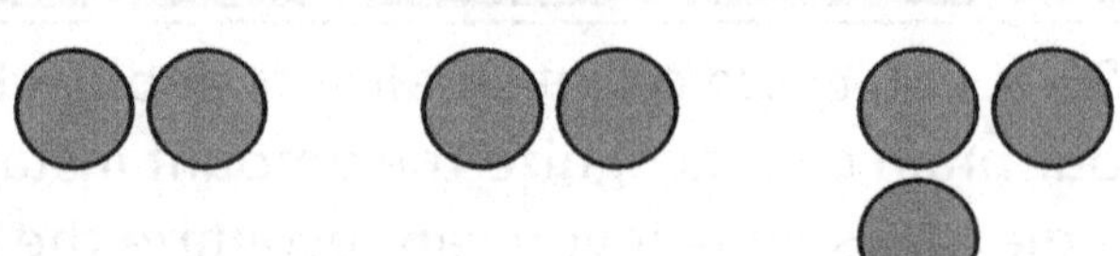

"Recognizing Amounts" Routine

How many do you see? How do you see them?

We all see amounts in a different way—even adults! Just look at this dot pattern:

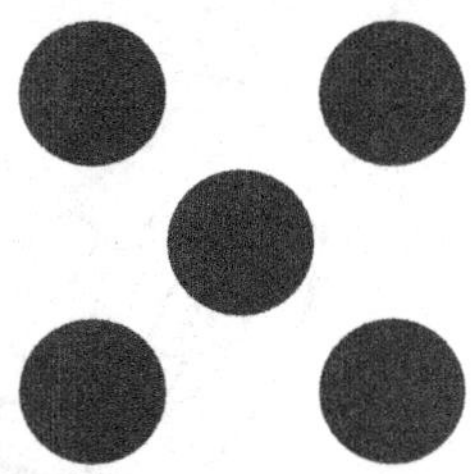

Some people see the 4 corners first, then 1 dot in the middle...

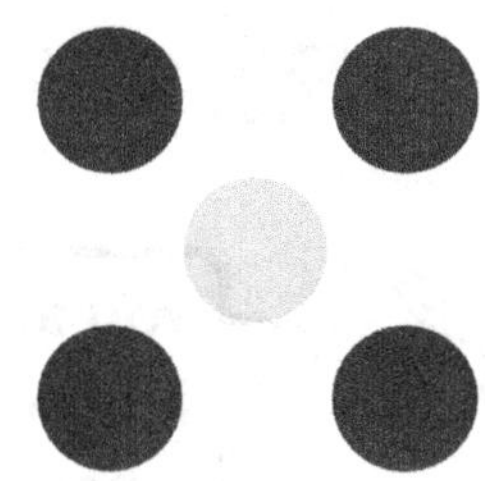

...and other people see a line of 3 dots first, then the dots on the top and bottom.

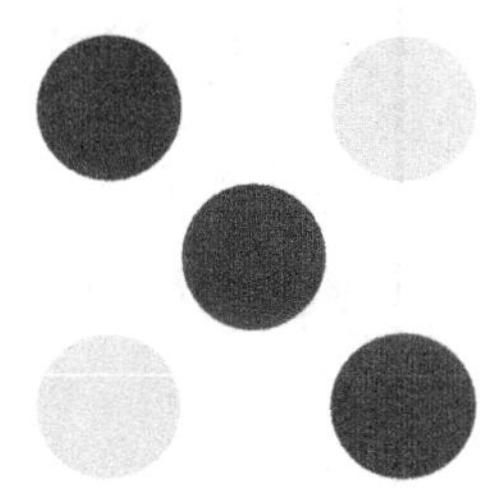

This becomes even more relevant when the numbers get bigger. For larger amounts like this, we don't automatically see the total. Instead, we quickly see smaller groups and add them in our head.

Scattered objects have many potential groups

Familiar patterns (like dice) highlight certain groupings

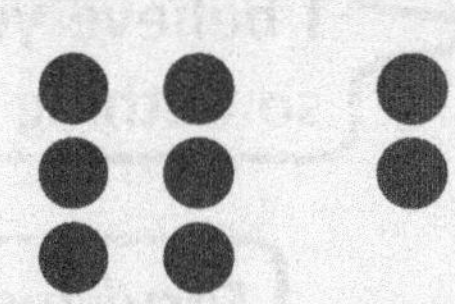

This helps me start to see groups of amounts. I bet that will be useful when numbers get a little harder!

Zero Alert!

That must be everything we can say about counting, I can't think of anything...

Don't forget about me!!!!

You are right! Zero is a *very* special number that too many people ignore. Zero is easy to forget because we can't see it, but knowing what it means can help with lots of math ideas later on:

Place Value	106 = One "hundred", **no** "tens", and six "ones"
Addition	6 + 0 means I add **nothing** to 6
Multiplication	0 x 5 means there are **no** groups of 5, which means zero
Fractions and Algebra	0 / *n* It doesn't matter what *n* is...**no** pieces of something is zero

I believe you, zero IS important! But how do I teach something that's invisible?

Easy! Just remember to ask questions about "no" objects. Count down all the way to zero as you take objects away to show that "zero" means "none."

When your daughter realizes she has ZERO fruit snacks left, she will remember what that means!

Changing the Difficulty (Arrangements)

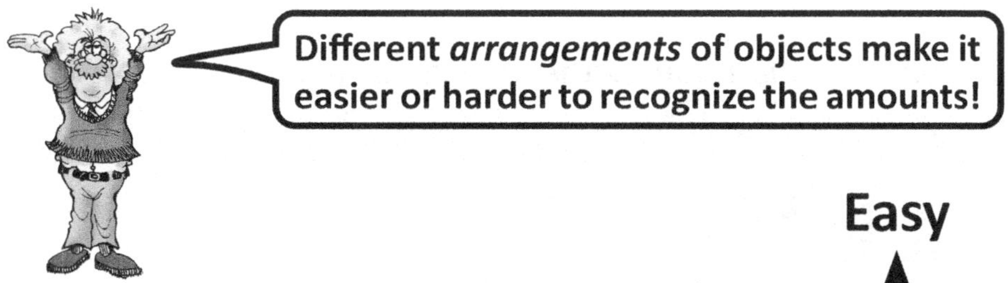

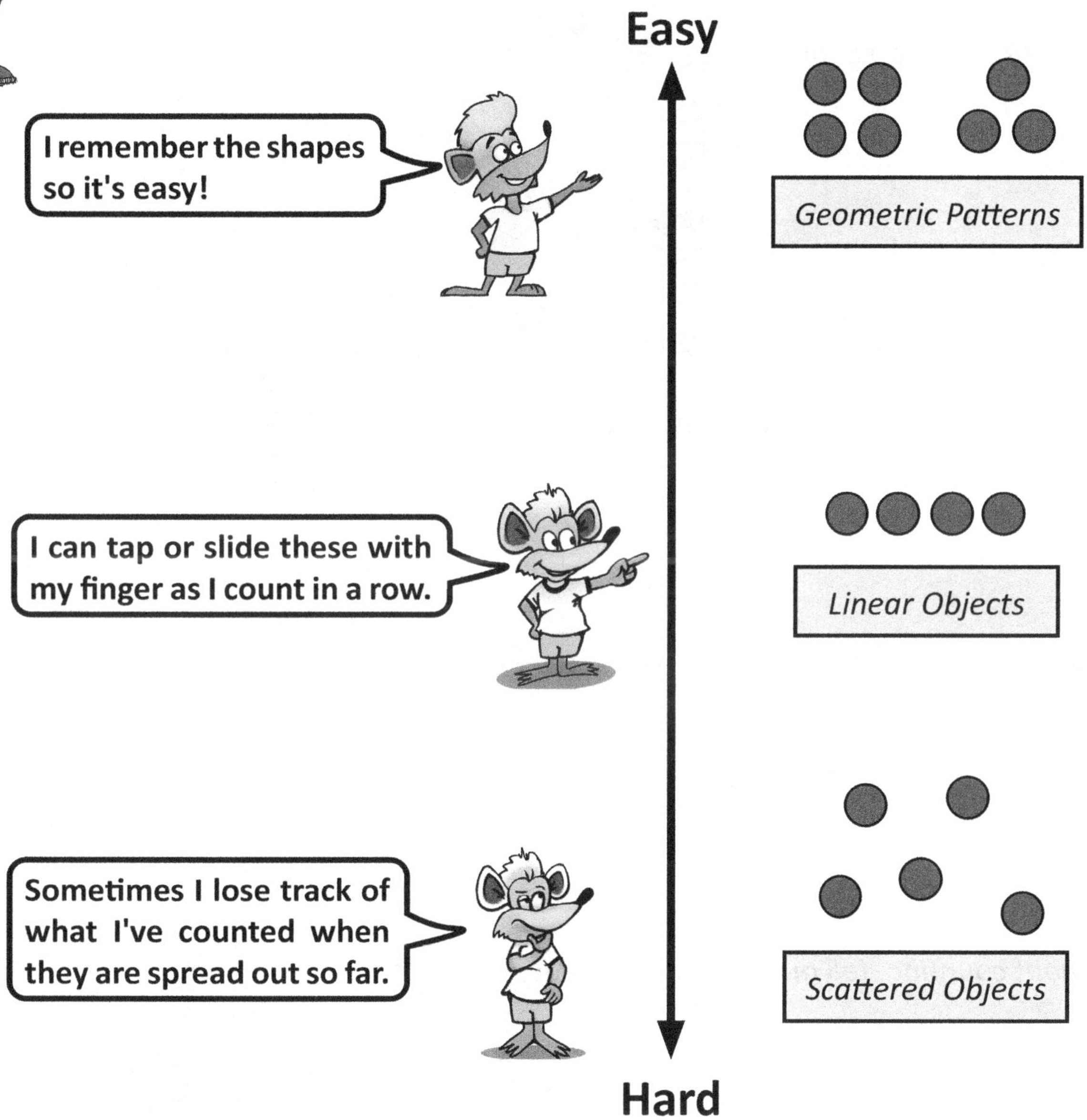

Automatically recognizing small amounts happens naturally as we grow. Children cannot be *forced* to do this if they are too young. It's good to model and practice this idea, but take a break and come back to it later if your child gets frustrated.

(See page 332 for dot cards you can print and use in your routines.)

Recognizing Small Amounts

Level 1

If a child insists on counting for these questions, show the objects for a second before covering them up and asking again.

1) *How many eyes do people have?*

2) *Without counting, tell me how many apples are here:*

3) (Hold up 1, 2, or 3 fingers.) *How many fingers am I holding up?*

4) *Without counting, tell me how many cookies are here.*

5) Which group of oranges is different?

Recognizing Small Amounts

Level 2

1) *Draw one circle.*

2) *Which group of apples is different? What makes it different?*

3) (Hold up 5 fingers.) *How many fingers am I holding up?*

4) *Without counting, tell me how many cookies are here.*

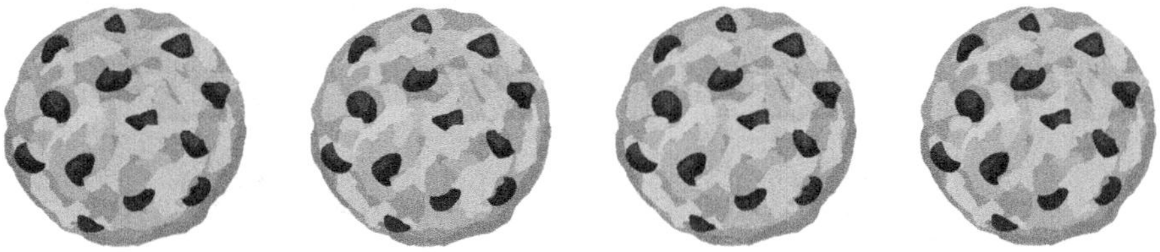

5) *How many crackers are on the plate?*

Recognizing Small Amounts

Level 3

1) *How many bears am I holding in my hand?*

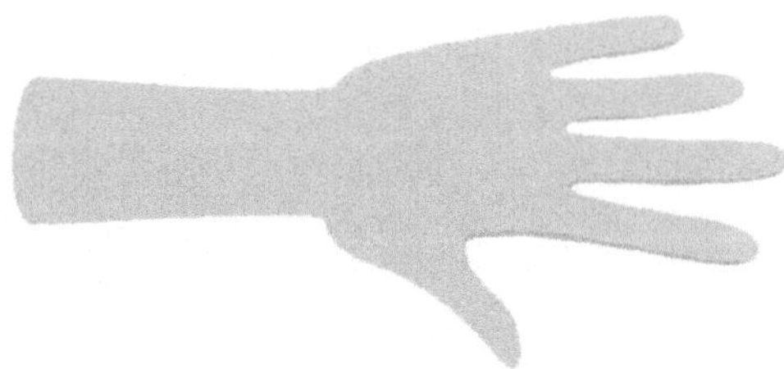

2) (Show these apples for 2 seconds, then cover them.) *How many apples were there?*

3) *Which group of crackers is different? How is it different?*

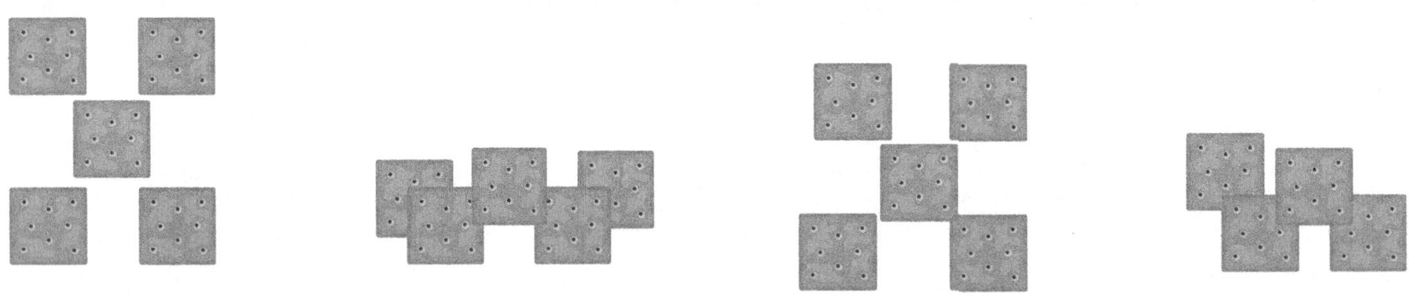

4) *Which group of dots is different? What is different about it?*

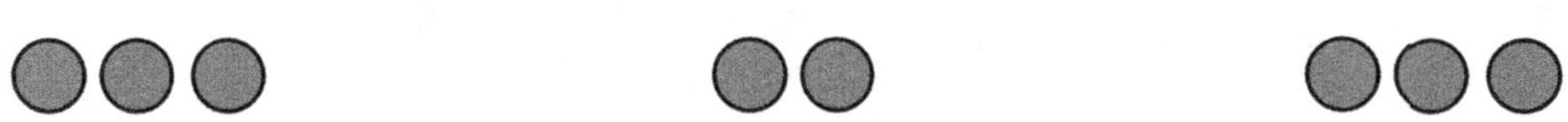

5) (Show these oranges for 2 seconds, then cover them with your hand.) *How many oranges were there?*

Recognizing Small Amounts

Genius Level

These are advanced questions and might be too hard for young learners (and that's ok)!

1) (Show these dots for 2 seconds, then cover them with your hand.) *How many dots were there?*

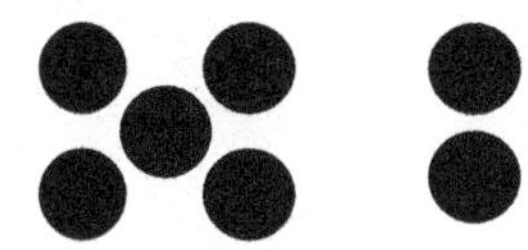

2) (Show these dots for 1 second, then cover them with your hand.) *How many dots were there?*

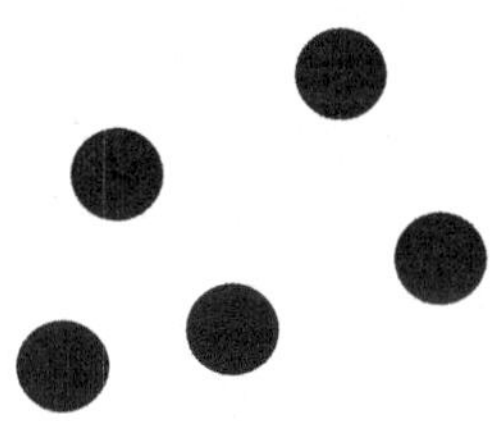

3) *Which group is bigger? By how many dots is it bigger?*

4) (Show these dots for 2 seconds, then cover them with your hand.) *How many dots were there?*

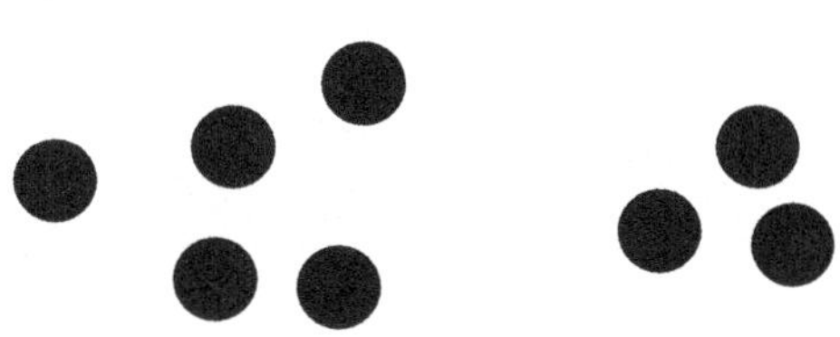

5) *Which group is bigger? By how many dots is it bigger?*

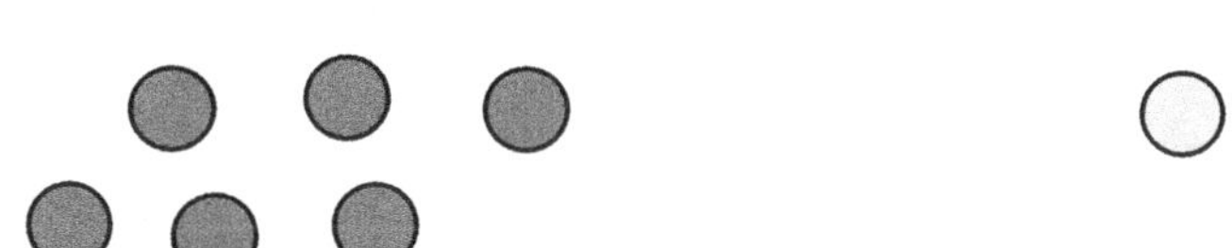

9

Counting Out Loud

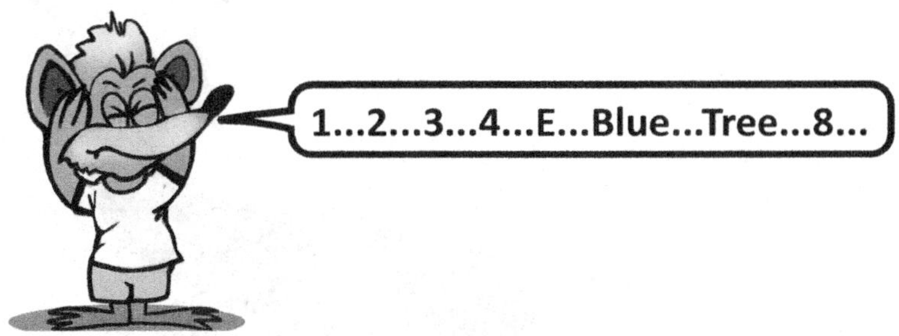

Children can count out loud before they have any idea what the numbers actually mean. To them, "one, two, three" might as well be "chair, lamp, floor," because they don't have a connection between the numbers yet. This worked the same way when we learned the alphabet song, and it probably took a while to realize that ELEMENOHPEE was actually L, M, N, O, P.

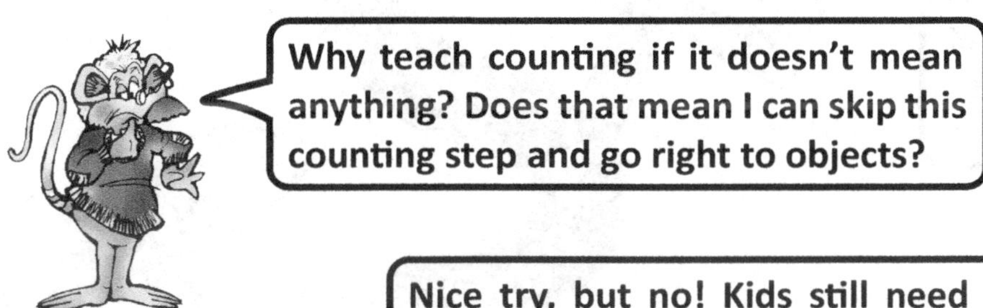

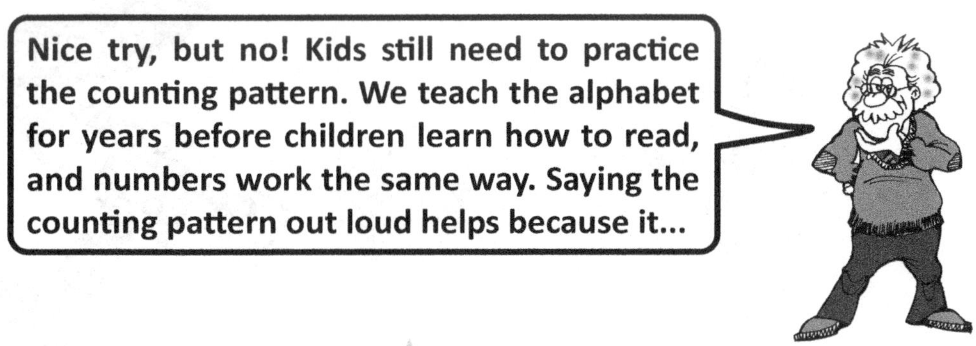

1. **Shows that numbers are important**

2. **Leads to correct *object* counting**

3. **Connects with important ideas like "one more" and "one less"**

Road Check

Ok, I believe you. But I'm a little lost on where this fits in with the big picture.
Let's take another look at Number Sense Road to see where we're going!
Number Sense
Join and Separate
Count for "How Many"
Count Objects
Number-Sense-Road
Count Out Loud
Recognize Amounts
Sorting and More/Less

This early step of counting is easy to understand but can take a long time to sink in. All you need to do is:

COUNT OUT LOUD!

That sounds boring if we are just standing around saying numbers all day. I'm going to count with some activities to make it more fun...

Washing hands

Brushing Teeth

"1, 2, 3..."

Eating Snacks

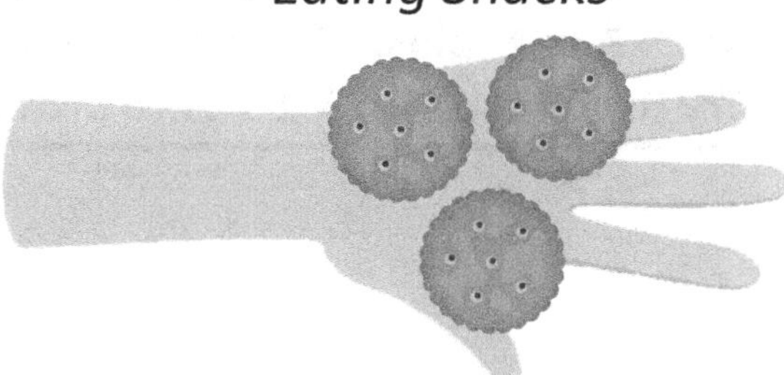

Playing Music

Ten is a very, very, VERY special number, so emphasize it every time! Say it louder, wave your hands, give a hug...anything to make **TEN** stand out to the child.

You don't have to count all the way to 10 every time, but it is important to hear the whole sequence often.

Ready to mix things up? Change the pattern every now and then to build flexibility as children learn more:

- **Touch objects as you count out loud**
- **Start counting from a different number (3...4...5...6)**
- **Count all the way to 20**
- **Skip count by 2s or 5s (2...4...6...8...10)**
- **Count backwards down to zero**

It's fun to count down 3-2-1-BLAST OFF, but "blast off" isn't a number. "Zero" is, and an important one! Try to count down 3-2-1-Zero instead.

Can I still say blast-off after zero?

Troubleshooting Counting

My 3-year-old knows all her numbers but says them in the wrong order! Is she bad at math?

Actually, that happens a lot when children are starting to count! Keep modeling the correct order for her, and practice saying the numbers together (or have her repeat them after you).

Don't correct every mistake, just model the correct order.

Ok, now she says them in the right order, but she combines some of them together instead of treating them like separate numbers. She goes: 1....2.....345.....6.....789.....10.

That also happens when we learn the alphabet (think "LMNOP"). Try having her count at a much *slower* speed.

Changing the rhythm of counting helps break old habits and shows that each number stands on its own.

Remember that counting out loud is just a small step to number sense. For now we are just practicing the order of numbers so we can move on to counting objects when we are ready.

Counting Out loud

Level 1

After the child has been exposed to the counting pattern, use these questions for extra practice. You can adjust the numbers to make them harder or easier. If any questions are too hard, model the skill throughout the day before asking again.

1) *Count* ***with me*** *to five.*

2) *Count* ***with me*** *to five while we...* (Do some activity for each number, examples below.)

   - Play notes on a keyboard (ascending)
   - Jump forward
   - Play a toy drum
   - Clap

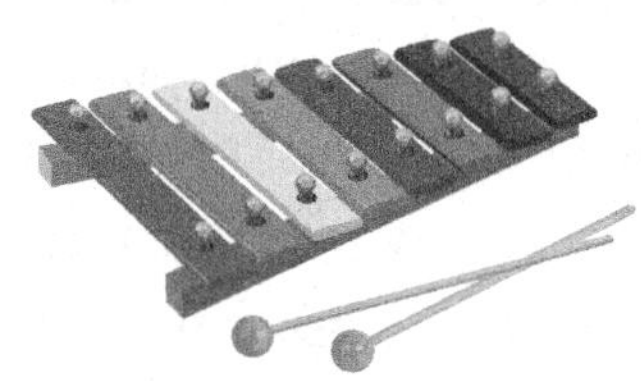

3) *Count* ***with me*** *to ten.*

4) *Count from 1 to 5 slowly while* ***I*** *touch these blocks.*

5) (Get 5 crackers or similar snack). *Count each cracker as I hand them to you...*

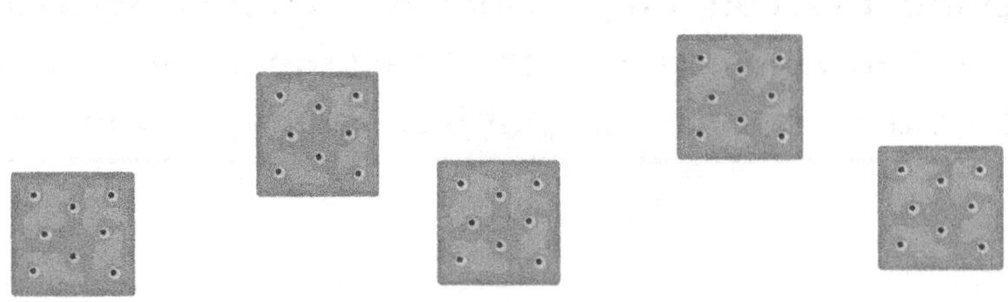

6) *Count to five.*

7) *Count to five while **you**...* (Do some activity for each number, examples below.)

- Play notes on a keyboard (ascending)
- Jump forward
- Play a toy drum
- Clap

8) *Count to ten.*

9) *Touch each cat as you count them.*

10) (Put 7 crackers or similar snack in front of the child.) *Slide each cracker as you count them...*

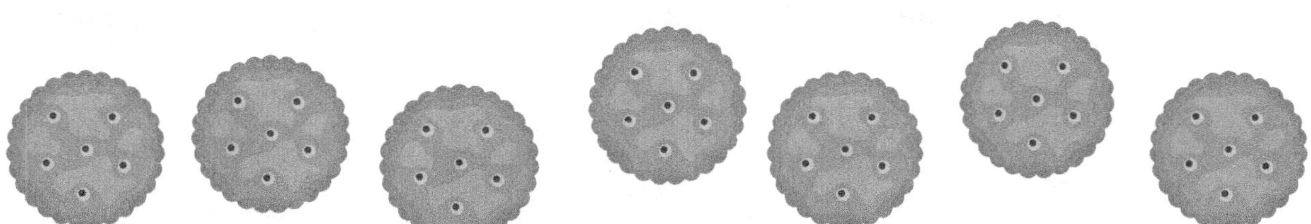

Counting Out loud

Level 2

1) *Count down **with me** from five to zero.*

2) *Count down with me from five to zero while we...* (Do an activity for each number.)

   - Play notes on a keyboard (descending)
   - Jump forward
   - Play a toy drum
   - Clap

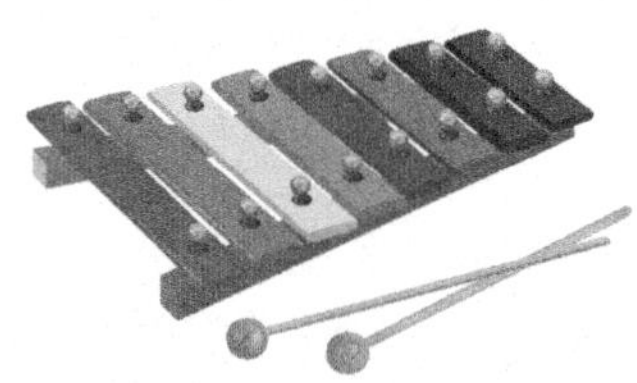

3) *Count down with me from ten to zero.*

4) (Hand the child 5 crackers or similar snack and take them away one at a time.)
   Count backwards from five to zero.

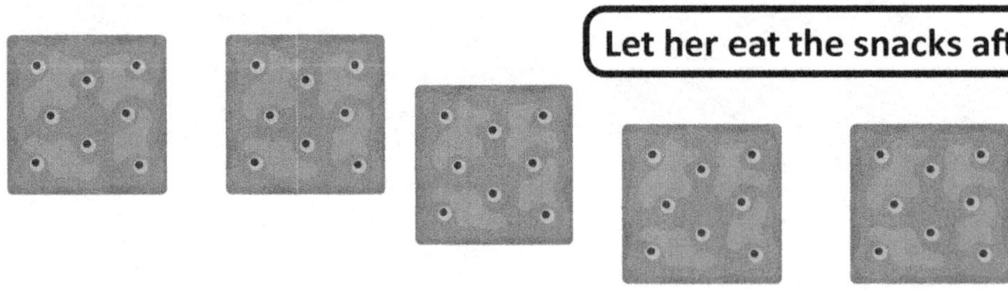

5) (Put ten pennies in a box and take them out one at a time.) *Count with me from ten to zero.*

6) *I'm going to start counting, then you finish counting to 10.* (Count from 1 to 6, then stop and let the child finish.)

7) *I'm going to start counting backwards, then you finish counting down to zero.* (Count down from 7 to 4, then stop and let the child finish.)

8) *Count backwards from 10 to zero.*

9) (Roll a die.) *Start at __ and count to ten.*

10) (Roll a die.) *Start at __ and count down to zero.*

Counting Out loud

Level 3

1) *Count from 1 to 20.*

2) *Count down from 10 to 3, and then back up to 10.*

3) *Start at 13 and count to 20.*

4) *Start at 17 and count to 25.*

5) *Count these cookies **without** touching them.*

Counting Out loud

Genius Level

Some of these questions go beyond the normal verbal counting pattern. They might be too difficult for now, and that's ok!

1) *Count out loud and touch only the circles.*

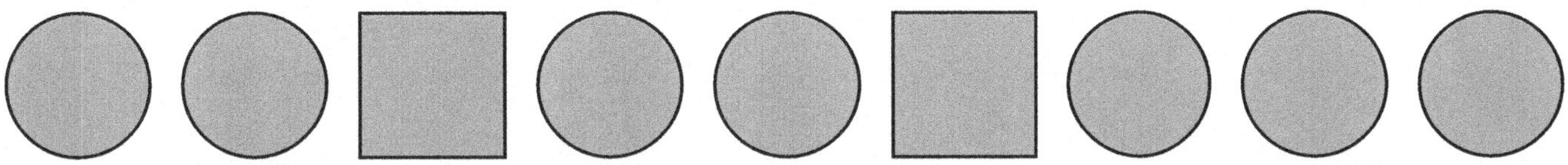

2) *Start at 16 and count backwards to zero.*

3) *Skip count by 2s all the way to 20.* (2, 4, 6, 8...)

4) *Skip count by 5s all the way to 50.* (5, 10, 15, 20...)

5) *Count backwards from 30 to 15.*

10

Counting Objects

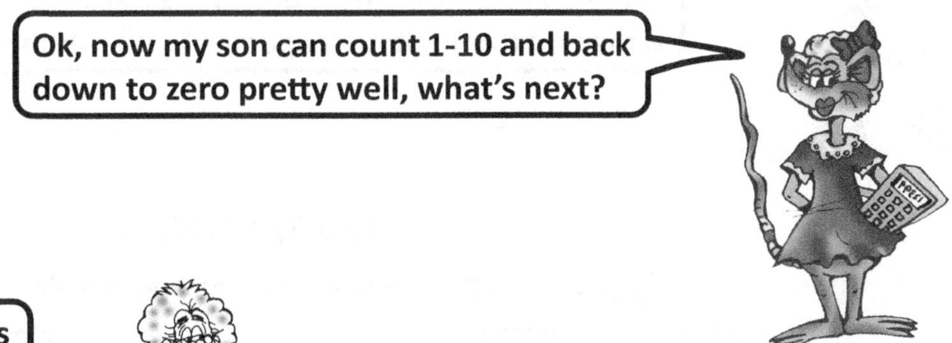

Learning the counting sequence is an important first step, but without connecting it to something *real,* we might as well be memorizing a list of ten random words. Our road to Number Sense tells us the next step is to emphasize the connection between the counting sequence and physical objects.

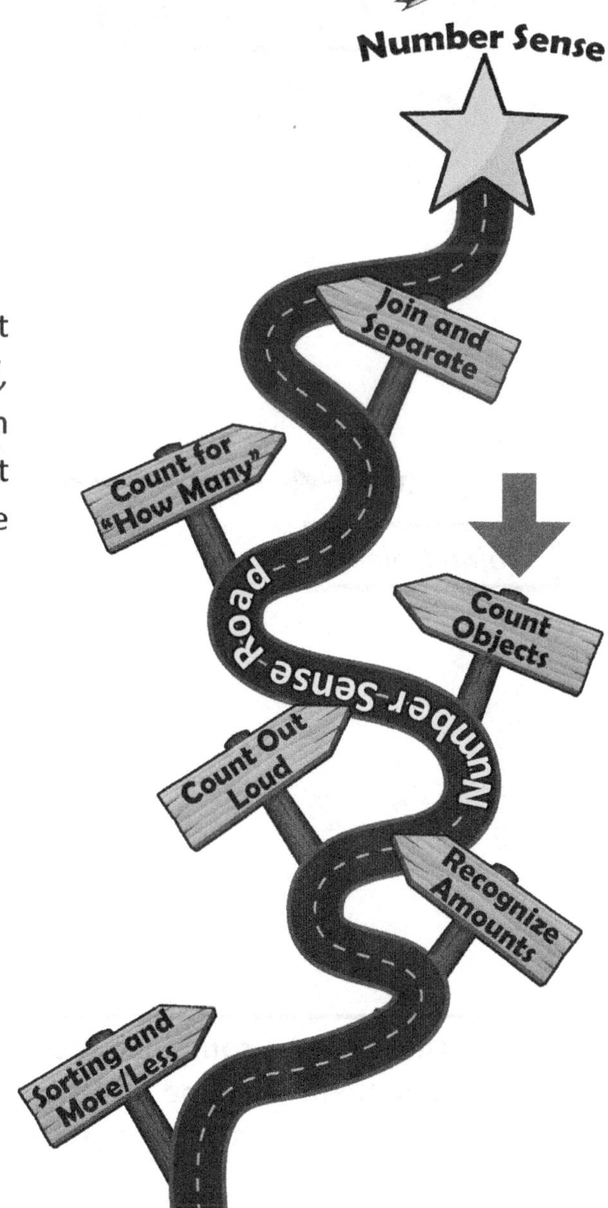

Arrangement vs. Interaction

We are NOT asking "how many" there are yet. We are just connecting the verbal counting pattern to real objects.

Some ways of counting objects are harder than others. Two things matter a lot: arrangement and interaction.

Arrangement

Easy ⟷ **Hard**

Objects in a line

Scattered Objects

Interaction

Easy ⟷ **Hard**

Sliding Objects

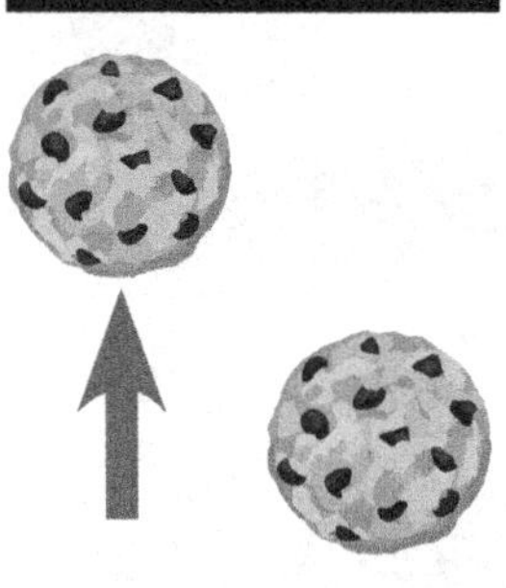

Tapping Objects

Seeing Objects

I bet these can be combined too! So the *easiest* way to count objects is to put them in a line and slide them as you count. The *hardest* would be to count scattered objects without touching them at all!

Ever try to do too many things at once? A child's brain can also get overloaded while counting! With large sets, we have to remember which objects we've already counted and which ones we haven't. This is harder to do if numbers are new and we still need brain power to think about what comes next.

Sliding objects as I count helps me keep track of which ones I've counted already!

I like counting things by nodding my head for each object. It's almost like tapping them with my finger—just a little harder!

What's harder than counting objects we touch? Counting objects we **don't** touch! This is hard at first, so one way to transition into the skill is to nod your head as you count (or do something else like tap the table). Tricks like this are like training wheels that help us move on to more complex skills. They aren't the end goal (we want to count with our eyes and mind), but can help get us there.

Hidden Objects

I still remember a hard question my dad asked me when I was a kid. He showed me 4 apples and said "there are 2 more apples under the bowl, how many total apples are there?" It was really hard to count apples I couldn't see, and that problem really made my brain steam!

Hidden objects are indeed the hardest to count!

Pairing

Here is one final idea you should know about counting. It might seem obvious to adults, but children need practice with it!

This is the idea that each object has a match. It helps transition kids to thinking about "amounts" and that we count to find "how many."

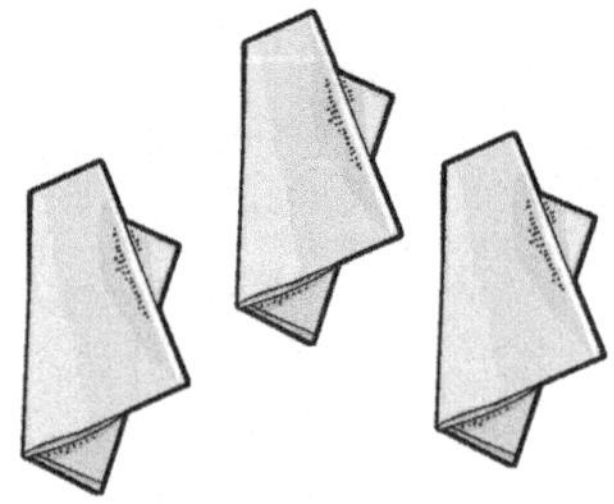

My parents helped me learn this by finding things that naturally go together (like setting the table by matching 3 napkins to 3 plates).

Counting Objects

Level 1

Use these questions as guides to practice object counting. Remember you can make them more difficult by: scattering objects, increasing the number of objects, counting without touching, or counting hidden objects.

1) Count the cookies.

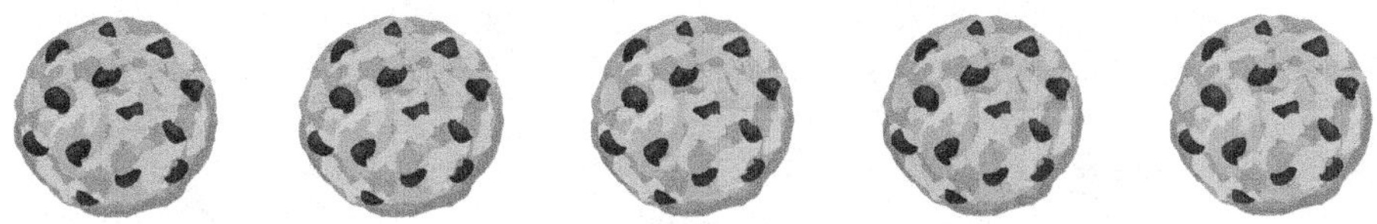

2) Count the bears.

3) Count the ants.

4) Count the rocks.

5) Count the apples **without** touching the pictures.

6) Count the fingers that are held up.

7) Count the dogs.

8) Draw a circle around every square, then count the number of circles.

9) (Grab a die and something to use as a game piece.) Roll the die and move the game piece along the board below:

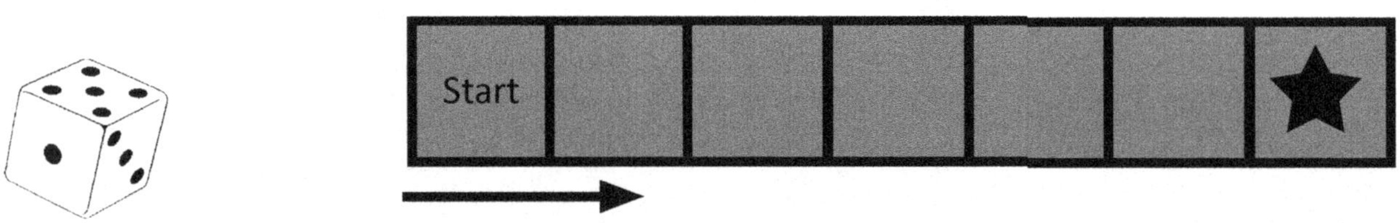

10) (Grab a die and something to use as a game piece.) Roll the die and move the game piece along the board below until you reach the end:

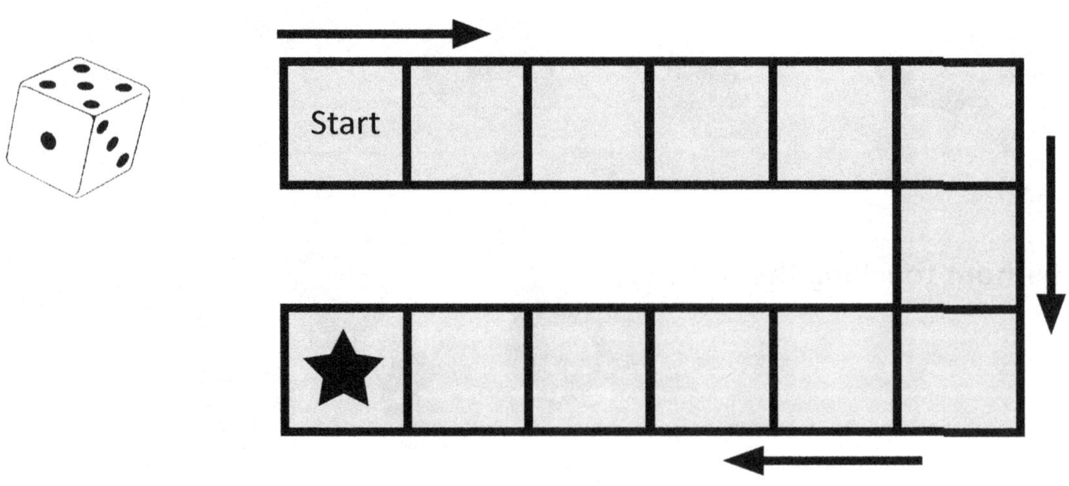

Counting Objects

Level 2

1) Put a checkmark in each dot as you count them.

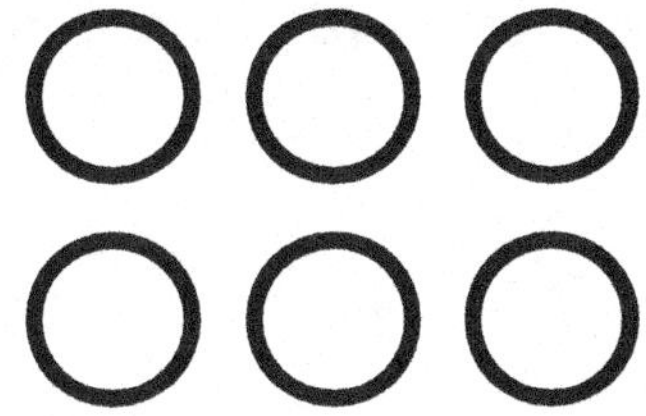

2) Count the bananas without touching them.

3) Count the squares.

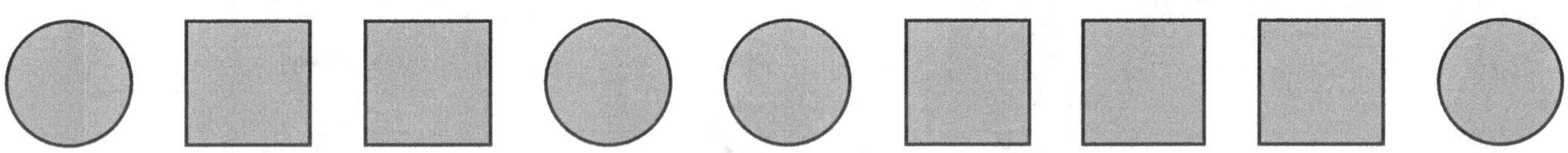

4) Count the bugs with wings.

5) There are 4 bears hiding under the bowl. How many bears are there altogether?

Counting Objects

Level 3

1) Count the black circles without touching them.

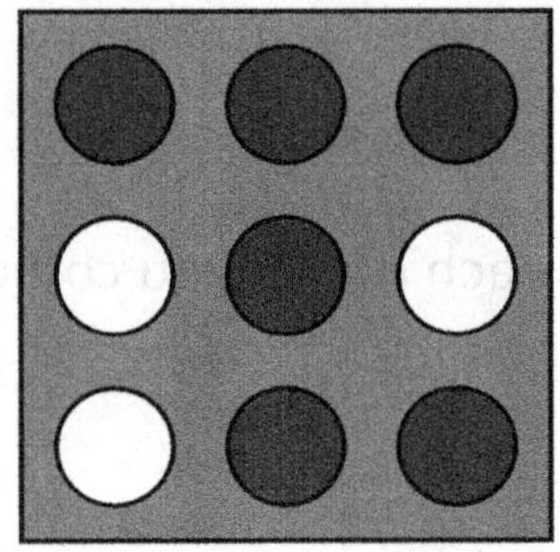

2) Draw more circles until there are a total of eight.

3) Count the bears, then circle the last number you said.

4 1

3 7

4) Count the cookies, then circle the last number you said.

1 5

7 2

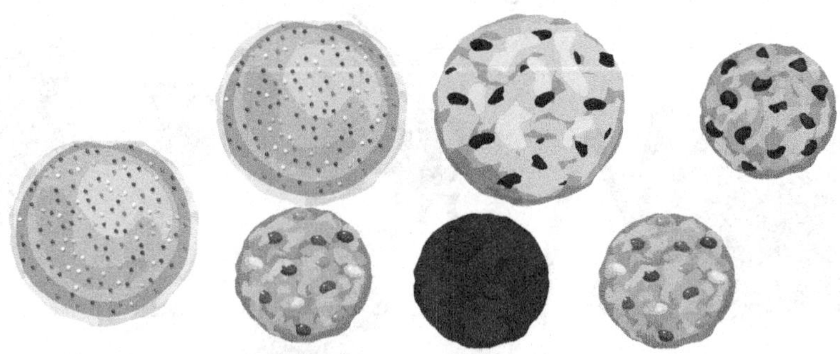

5) Five cats are hiding in the barn. Start at five, then count the rest of the cats to see how many the farm has in total.

Counting Objects

Genius Level

1) This is a number path. Count each box and write the missing numbers while you count.

1	2	___	___	5	6

2) Draw more circles until the amount matches the number.

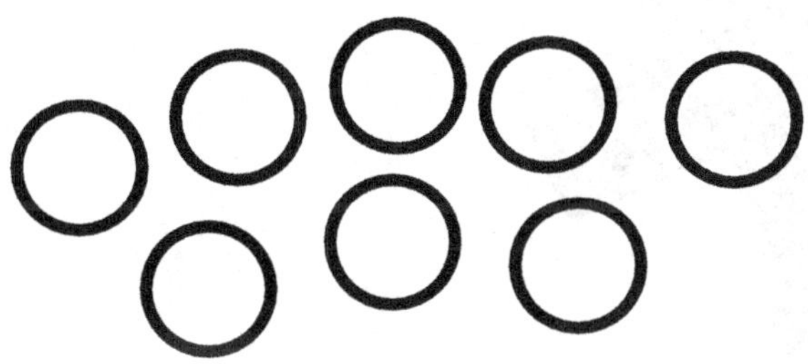

= 12

3) Count each group of objects out loud, then draw a line to the last number you said.

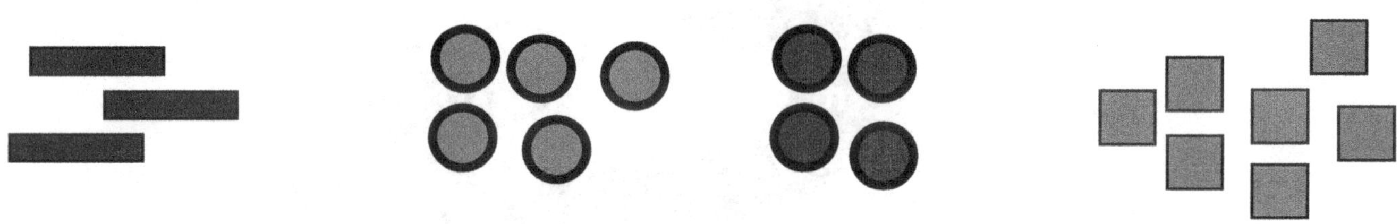

5 3 7 4

4) Count **every other** dog below. (Model this first if your child doesn't understand.)

5) There are three bears hiding under the bowl and seven bears hiding behind a curtain. Some bears forgot to hide. How many bears are there altogether?

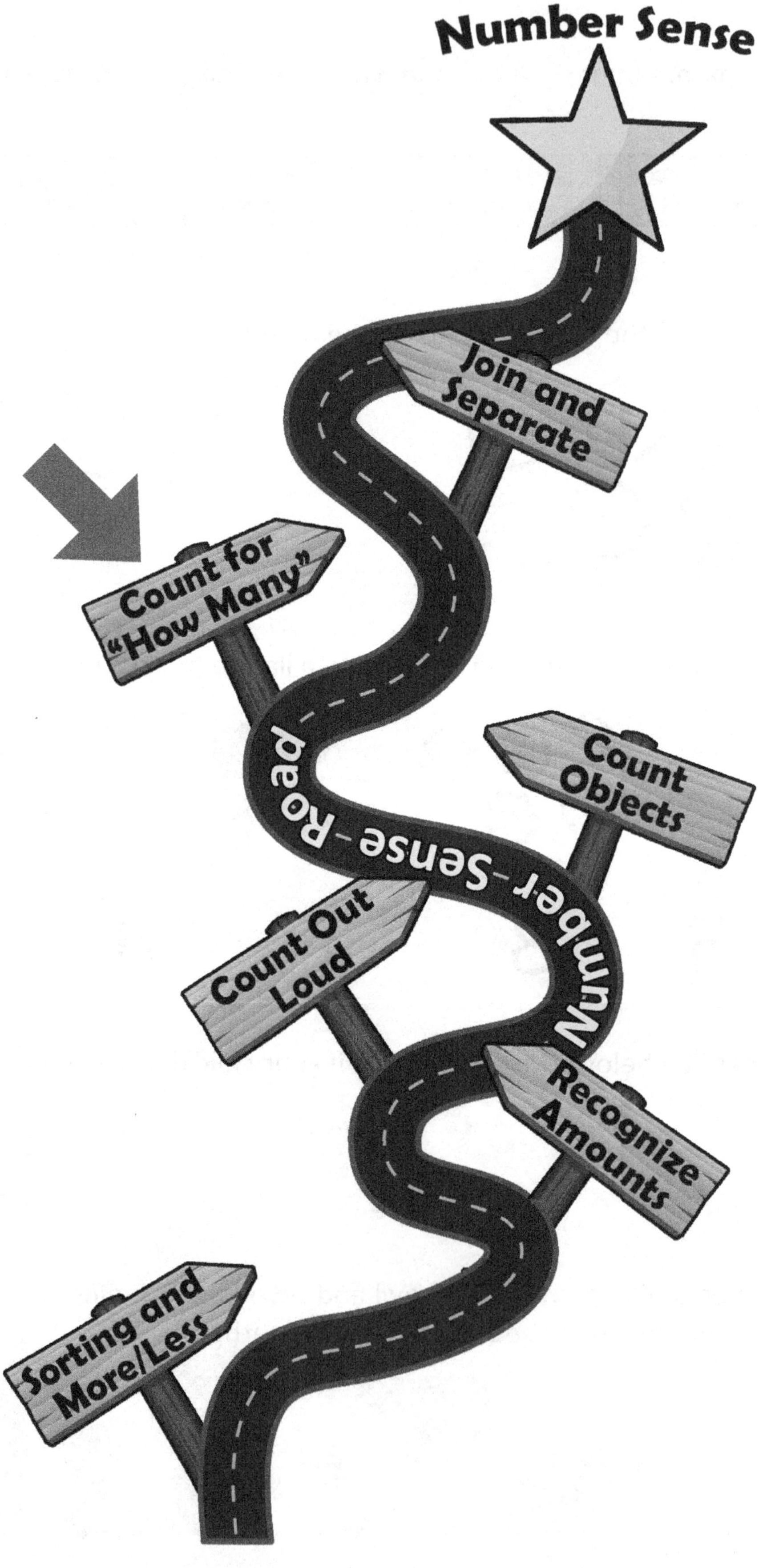
Number Sense
Join and Separate
Count for "How Many"
Count Objects
Number-Sense-Road
Count Out Loud
Recognize Amounts
Sorting and More/Less

11

Counting to Find How Many

This child thinks that "3" means the *last* orange he counted, not the whole group of them. He needs to learn that counting tells us "how many" we have. This means the **last counting number tells us the amount we have**.

This is a normal step in developing number sense, and can be baffling for adults. *"Of course we count to find how many we have...aren't we born knowing that?"* **Apparently not!**

Teaching "How Many"

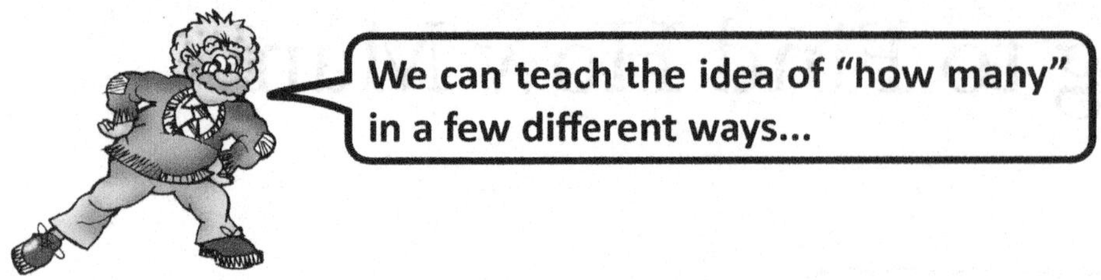

Show that the object order *doesn't matter* when we count

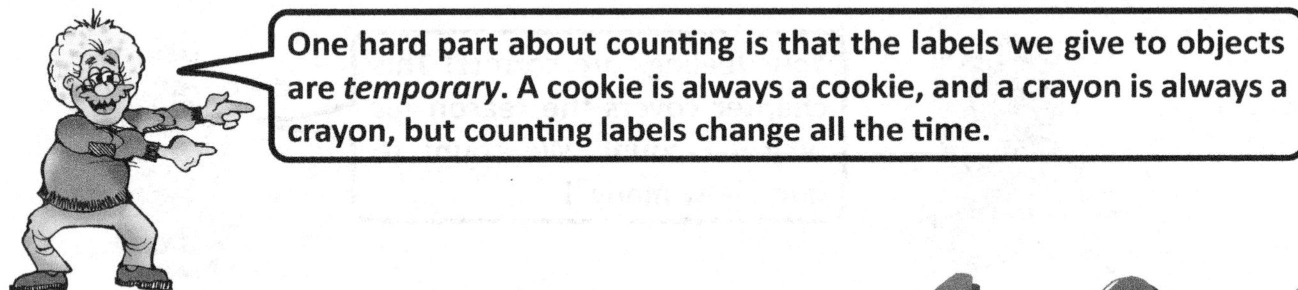

If we count to the right, the cookies are labeled like this...

BUT if we count to the left, those same cookies are labeled differently...

You can show that the counting order doesn't matter by starting at different sides when you count. Most people count left-to-right because that's how we read, but be sure to count right-to-left sometimes too.

Label small amounts

Label the amount after counting

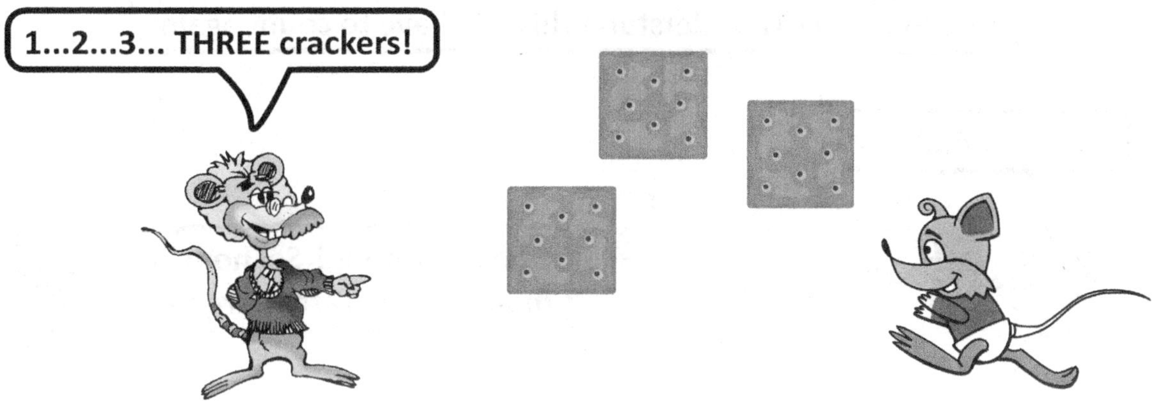

Show that the arrangement doesn't change the amount

Count, mix-up, and count again

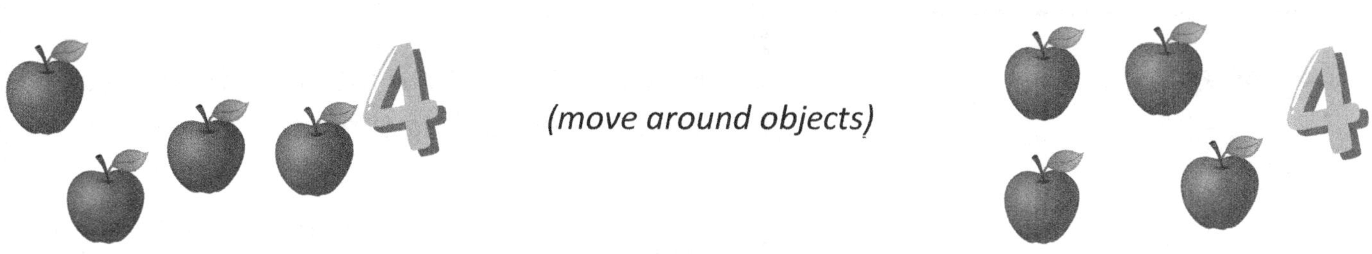

Spotting Success

- He knows the amount of objects stays the same even if they are mixed up.
- He can tell me "how many" cubes there are without recounting them.
- He can hand me six crackers when I ask him.

Counting to Find How Many

Level 1

If your child can't answer these questions (or needs to recount each time), you may need to model amounts more or wait for the child to develop further. Understanding "how many" isn't something we can force children to understand before they're ready.

1) Count the bears. How many bears are there? (Without recounting.)

2) Count the apples. How many apples are there? (Without recounting.)

3) Count the crackers. How many crackers are there? (Without recounting.)

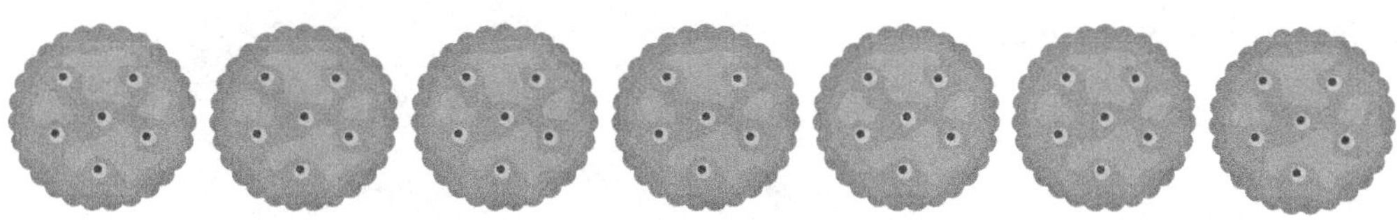

4) Count your fingers on one hand. How many fingers are there? (Without recounting.)

5) Count the oranges. How many oranges are there? (Without recounting.)

Counting to Find How Many

Level 2

1) Count the spiders. How many spiders are there? (Without recounting.)

2) Count the oranges **this way** ⟶ Now count them **this way** ⟵

3) How many tires does a car have?

4) Draw 7 circles and count them. How many circles are there?

5) How many grasshoppers are below?

Counting to Find How Many

Level 3

1) Count the squares from left to right. How many will there be if you count them the other way?

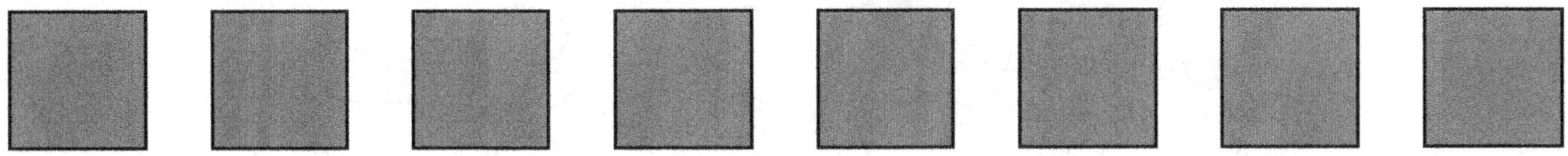

2) Draw 2 circles and 3 squares. How many shapes did you draw?

3) (Put eight crackers or other snack in front of the child.)
Say: Count these crackers.
(Shuffle the crackers around.)
Say: How many are there now? (Without counting again.)

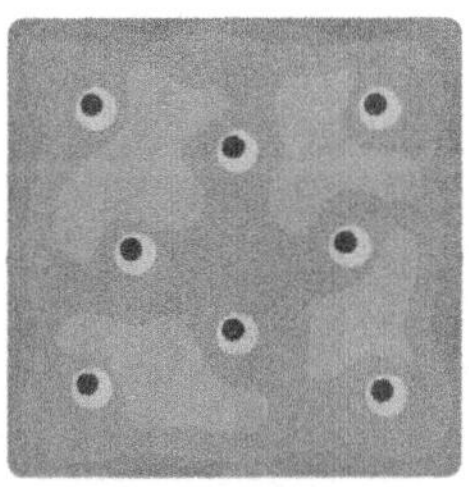

4) (Give the child a large pile of cubes or other objects.) Give me seven cubes.

5) Tell me three ways that these groups are different:

Counting to Find How Many

Genius Level

1) Each kid wants to eat a brownie. How many brownies do you need?

2) Draw 2 circles, 3 squares, and 4 triangles. How many shapes did you draw?

3) Point to the groups in order from **fewest cubes** to **most cubes**.

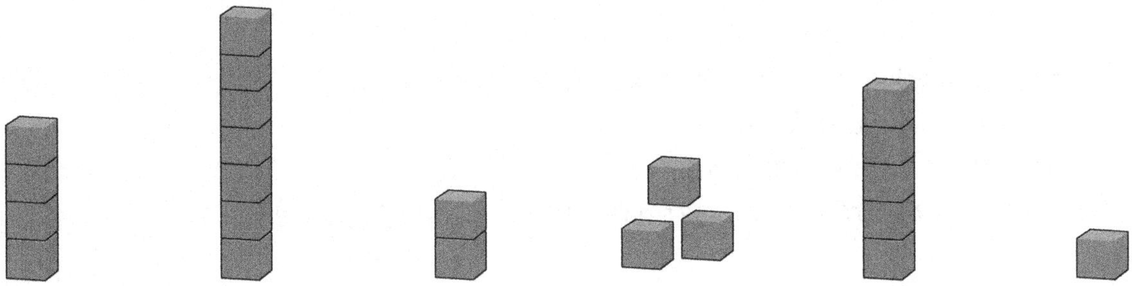

4) Which tank has more fish?

5) If each cat gets one toy, how many left over toys will there be?

12

Joining and Separating Numbers

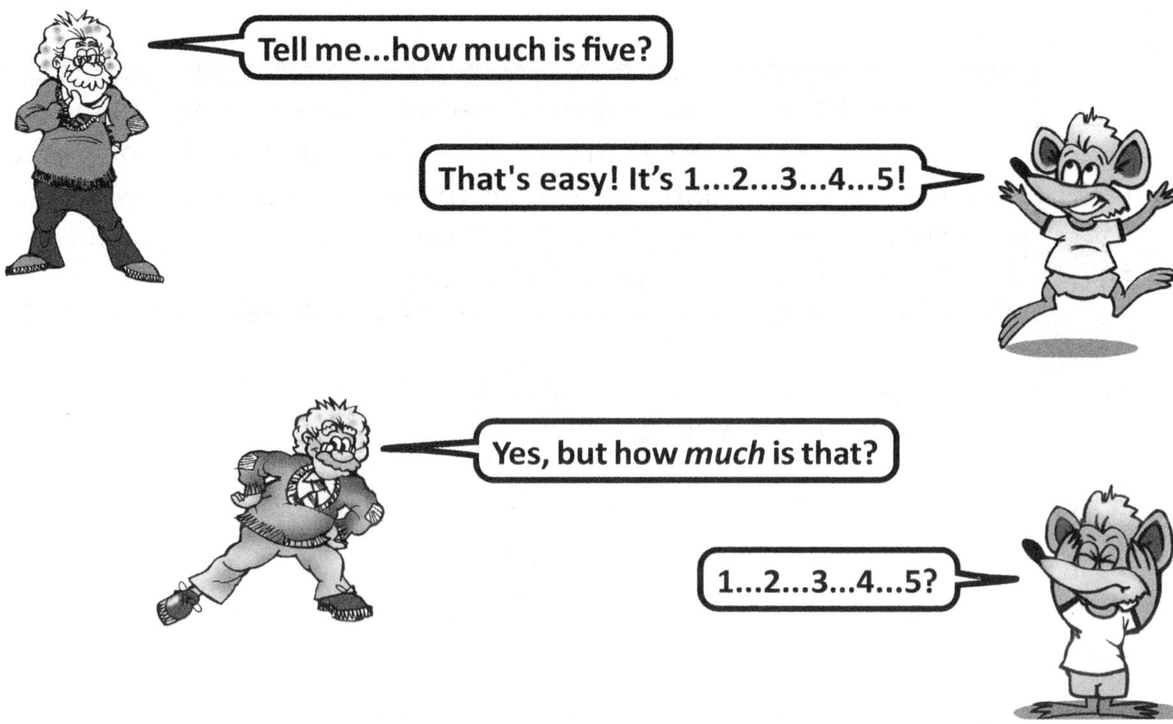

This budding mathematician is missing an important part of what numbers are. Every number has a special relationship to the ones before it. The order is *not* random.

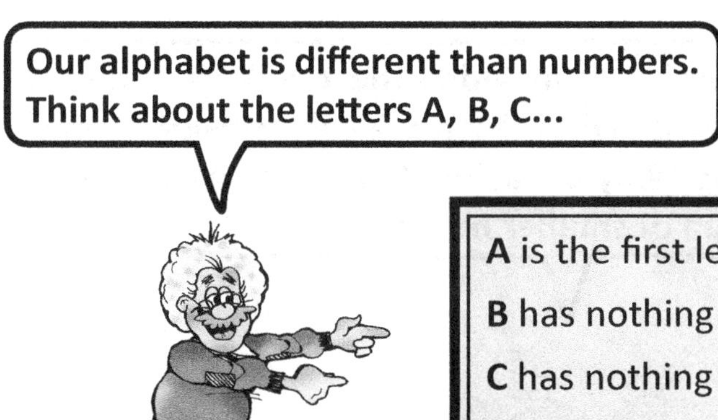

A is the first letter

B has nothing to do with **A**

C has nothing to do with **A** or **B**

Letters are separate symbols, they don't relate to each other at all!

Our alphabet only has a certain order because we all agreed to it. If we suddenly changed the alphabet song to *B-P-G-T-L-S-P,* it would sound weird but the way that we actually read would stay the same.

The first letters in our alphabet are not more common than the later ones. The most common letter in English is "S", but that doesn't show up until the end of the alphabet. That is why you should never teach letters in A-B-C order. A better approach is to start with the most common letters and move on from there.

How did you get so smart?

What makes our number system different than the alphabet is that *every number includes the number before it!*

The number 3 includes the numbers 2 and 1. This idea is called "Hierarchical Inclusion," which is a fancy way of saying that bigger numbers are made up of smaller numbers.

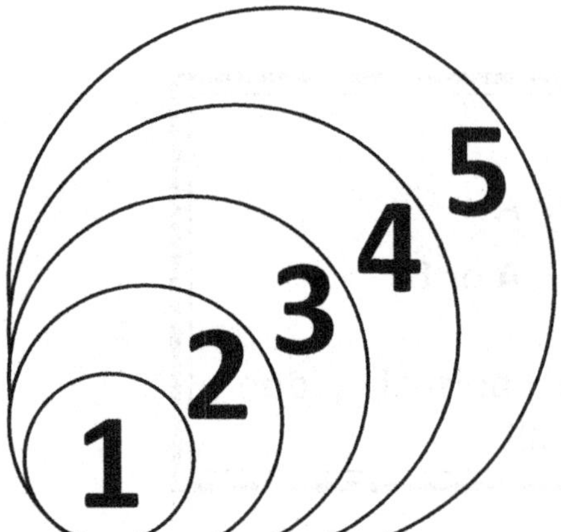

The last chapter showed us that we count to find "how many." It is also important to understand that numbers can be broken apart or combined to make new numbers. This sounds like "addition" and "subtraction," but don't use those fancy words or symbols yet! Kids learn math easier when they understand an *idea* before being told to memorize a *word*.

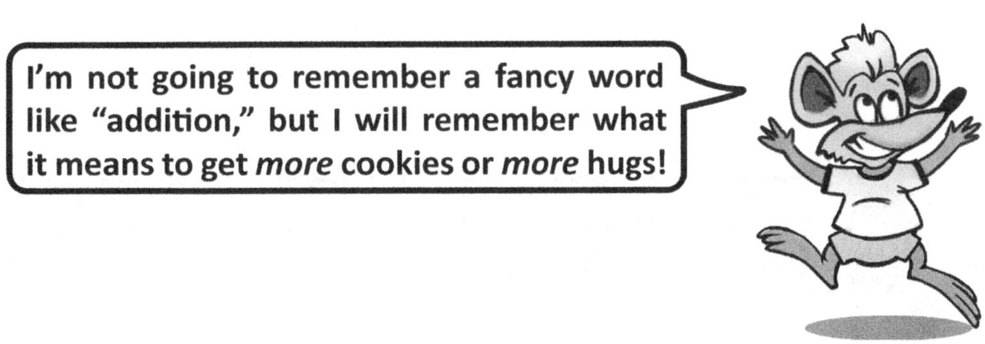

How Many Now?

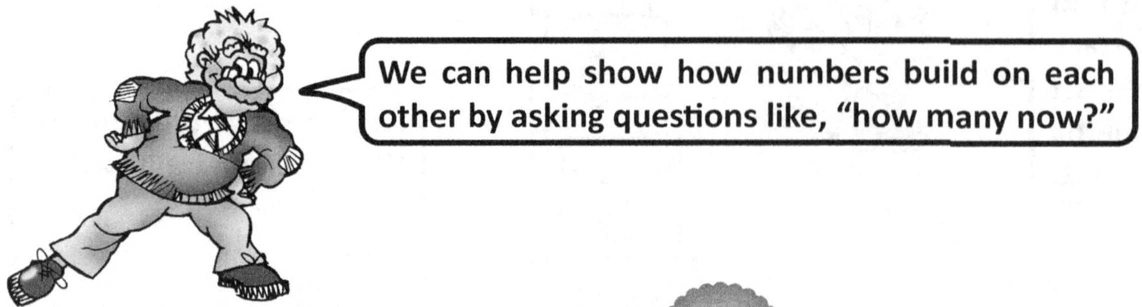

How many crackers do you have?

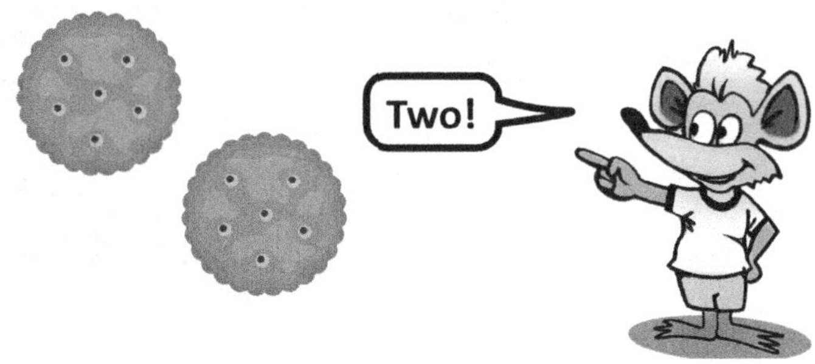

How many now?

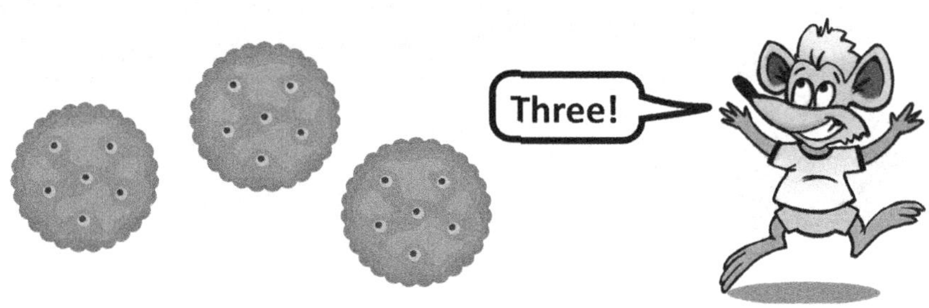

How many now?

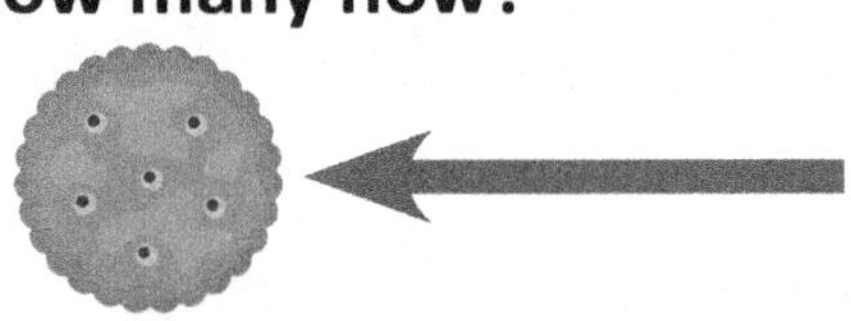

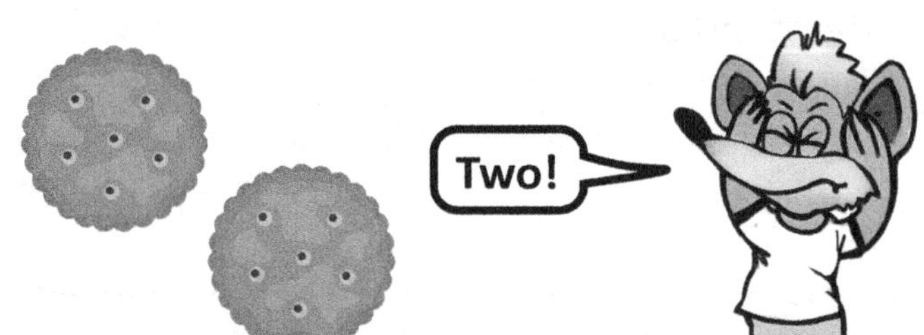

This makes *much* more sense with objects or drawings instead of numbers (1, 2, 3).

Four crackers shows an amount in a way that is easier to understand than the symbol "4". (Of course, this depends on the age of the children and how well they know their numbers.)

Abstract Dial

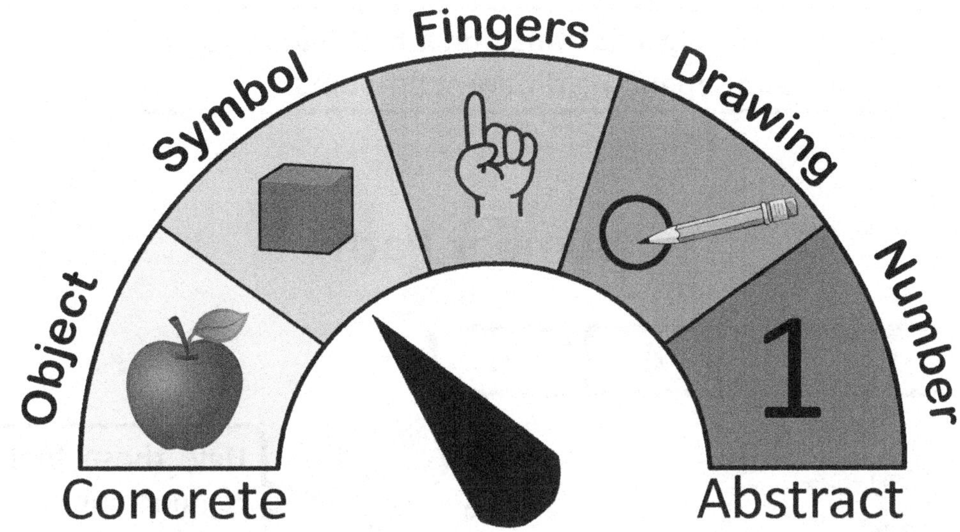

Remember the **Abstract Dial**! We can help children understand how to break apart numbers by starting with more concrete objects like cubes. They are a little harder to understand than something like snacks, but are still objects that children can hold and move.

Use different cube colors to make combining "sets" more visual!

3 cubes (blue) and

2 cubes (red)

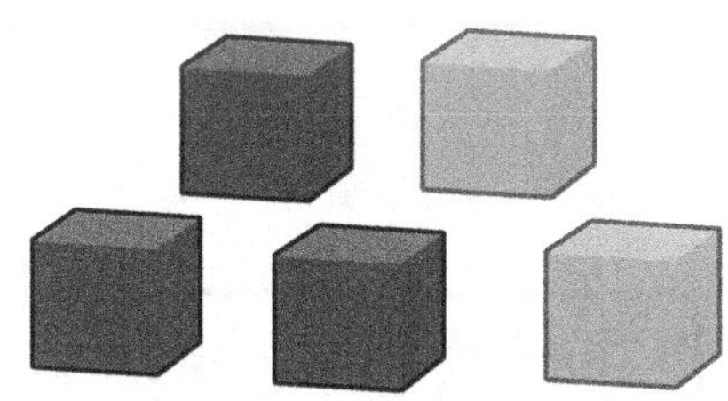

gives you 5 cubes altogether!

I like to go a little further and have cubes "represent" something.

"Pretend each cube is a cookie..."

Some children learn these concepts quickly and get bored with "baby" numbers. Smaller sets are easier to see and count, but it's also a good idea to try larger amounts to see what happens. (Just don't be afraid to make the numbers small again if you need to.)

Concrete Tools

Number Path

Number Line (distance)

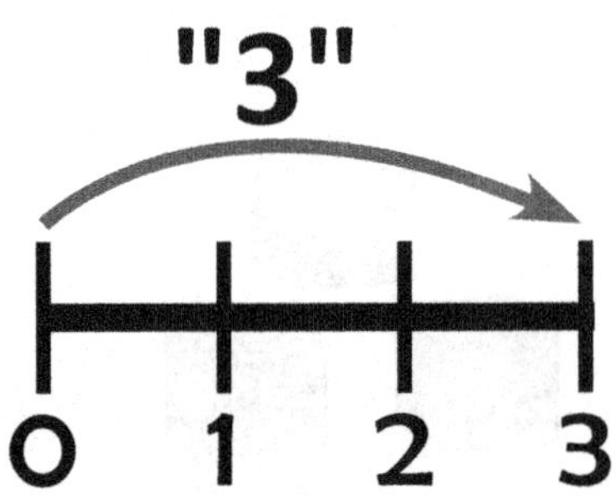

Number Path (objects)

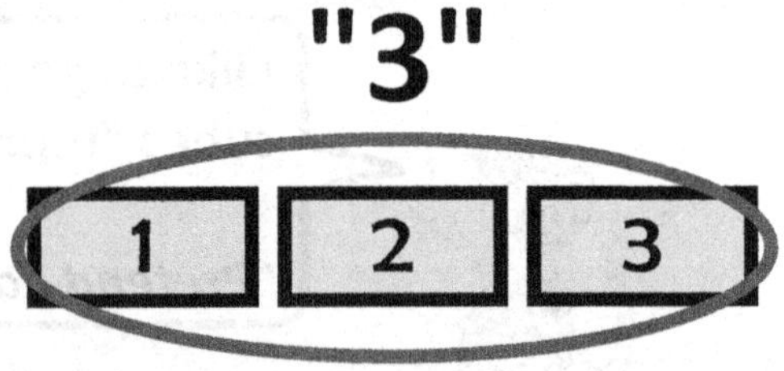

Number Paths can be used to count up, join, or separate amounts. Because each number is a separate object instead of a line, they are easier to group and think about.

(See page 329 for Number Lines and Number Paths that you can use.)

Rekenreks

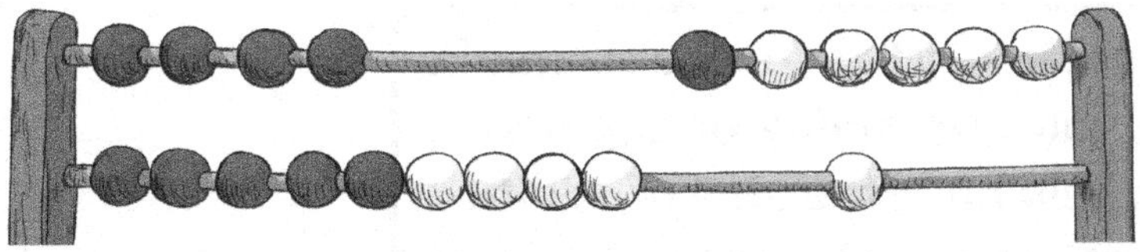

This is a Rekenrek. It is an excellent way to demonstrate joining or separating numbers (especially when talking about 5s and 10s).

The two colors can represent different groups you are combining, and children can physically slide the beads to show what happens to the amounts.

I can slide the beads as I count to 10! Maybe even for skip counting by 2s!

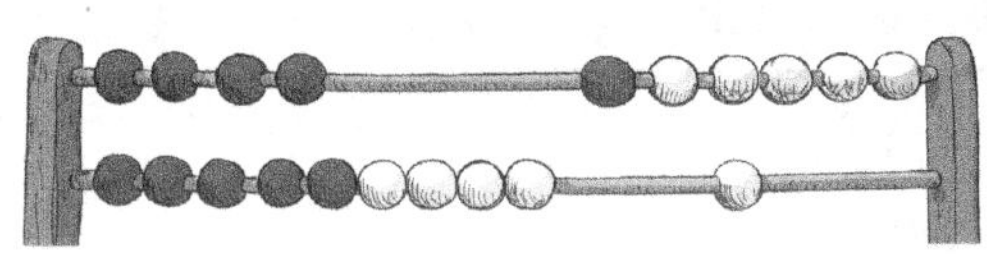

I can see combinations of 10 (like 4 and 6).

Or I could act out story problems to visualize what the question is asking.

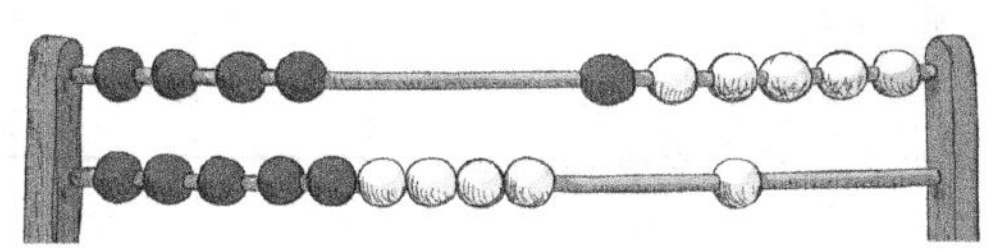

Rekenreks help show the importance of 5 and 10 as benchmarks for math. There are **10** beads in each row, and **5** of them are colored red.

Many stores sell these online, or you could make your own with pipe cleaners and beads!

Ten Frames

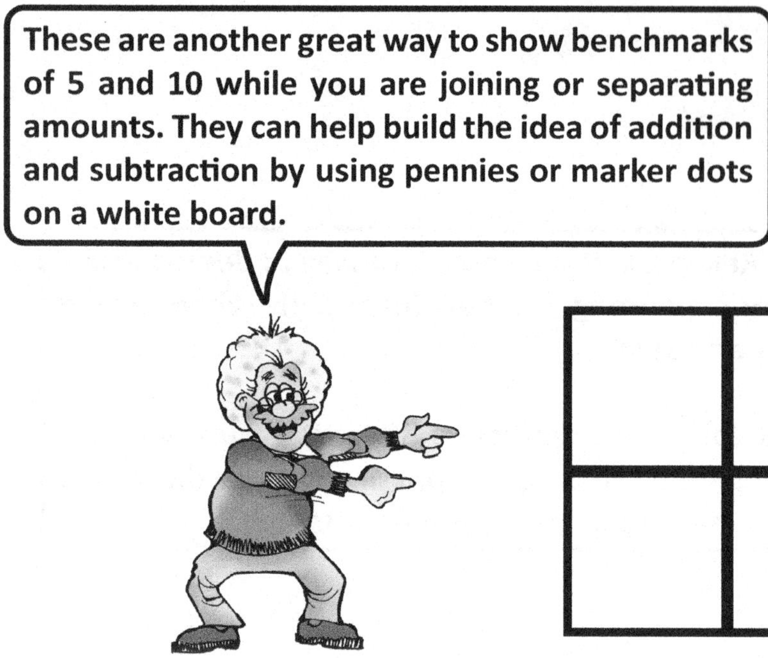

How many more to make 10?

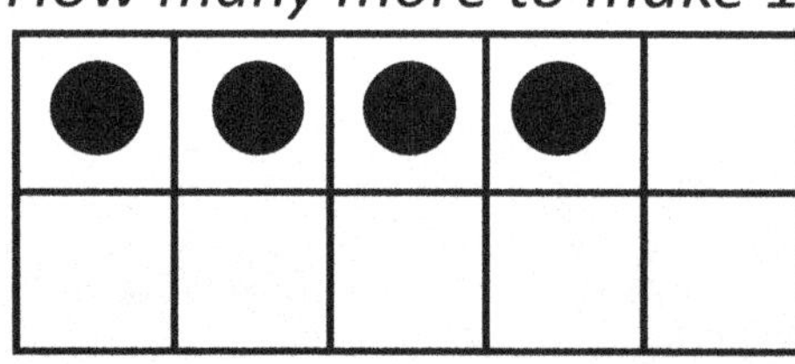

What's missing?

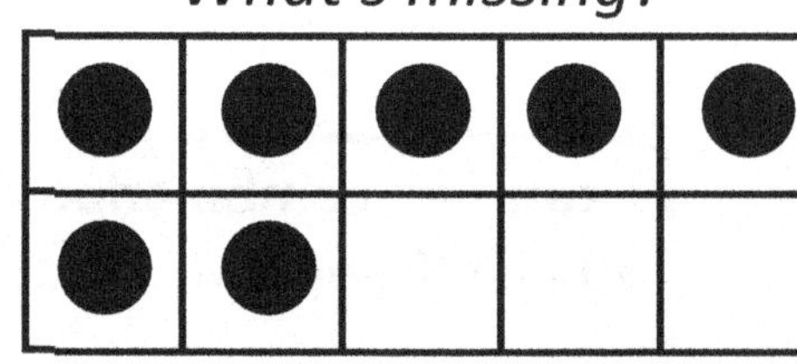

How many dots do you see?
How do you see them?

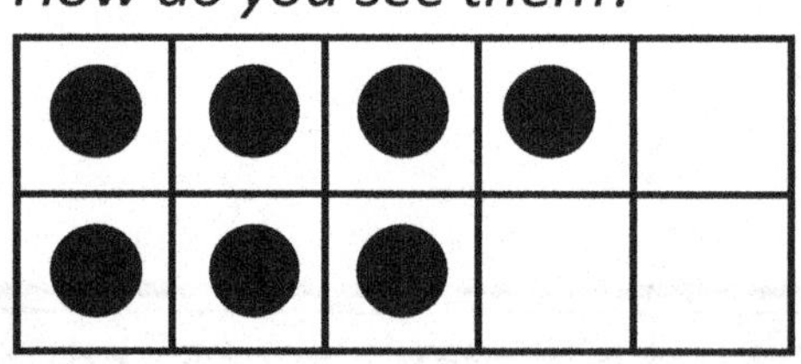

Add the rows

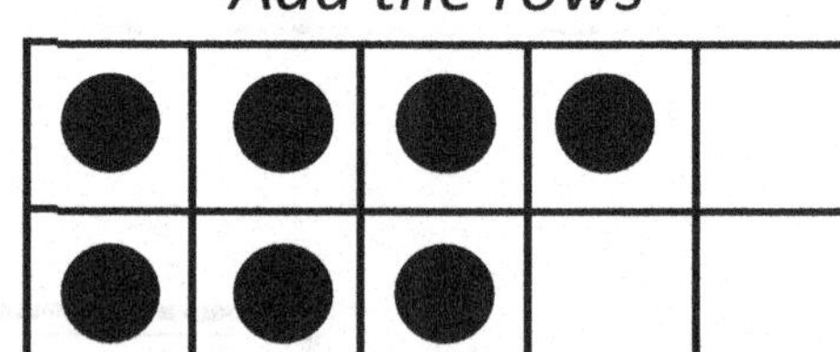

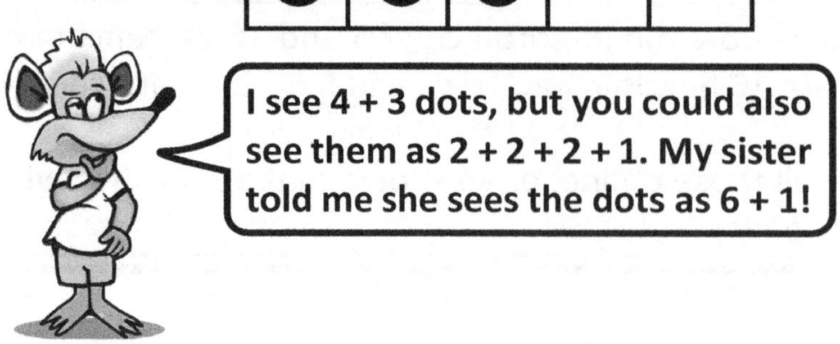

(See page 333 for blank Ten Frames you can use or print.)

Joining and Separating Numbers

Level 1

These questions are similar to "Addition and Subtraction" but focus more on the basic concepts of joining and taking away.

1) If you combine three crackers and two crackers, how many crackers do you have?

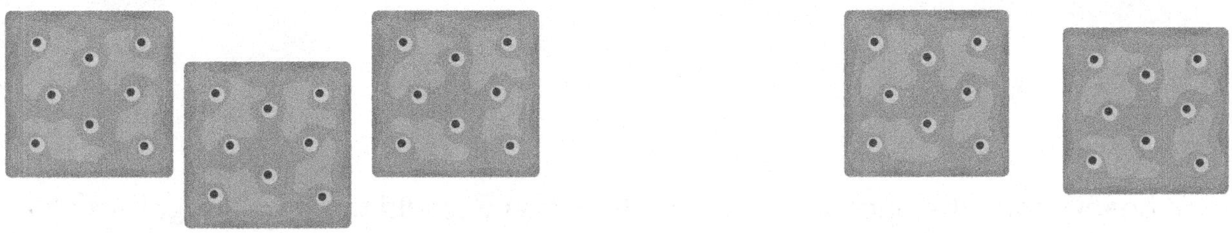

2) Two bears were sitting in a cave, then two more bears joined them. How many are there now?

3) How many **dice** are there?

4) How many **dots** do you see on these dice?

5) Jane had 6 cookies and her brother had 3 cookies. If they combined them, how many cookies would they have altogether?

Joining and Separating Numbers

Level 2

1) If you combine five crackers and five crackers, how many do you have altogether?

2) If three bears joined the ones shown below, how many would there be altogether? (Start with the bigger number and count three more.)

and 3 more = ?

3) My friend said she has three pets: some dogs and some cats. How many dogs and cats could she have?

4) Draw more circles until there are a total of seven shapes.

5) These children want to bring their cats for a tea party. If each person and cat wants to drink tea, how many total cups do they need?

Joining and Separating Numbers

Level 3

1) Draw eight shapes: some circles and some squares. What is another way you could draw them?

2) You are asked to draw 4 circles and then some squares. How many squares should you draw if you want a total of 8 shapes?

3) If you combine 5 crackers and 5 crackers, and then eat 2 crackers, how many do you have left?

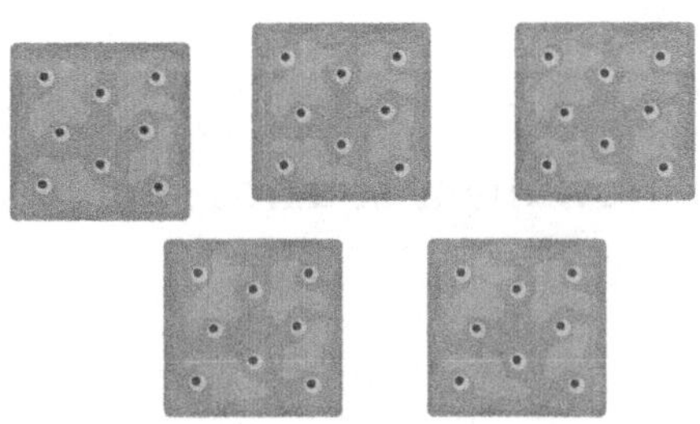

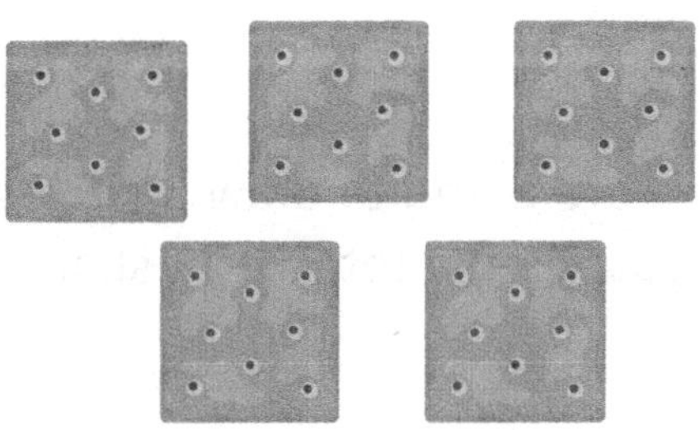

4) Two bears were sitting in a cave and then 3 more bears joined them. One bear was hungry and left to find food. How many bears are in the cave now?

5) Jenson and his brother each had 5 brownies. If they each ate one brownie, how many would they have left (total)?

Joining and Separating Numbers

Genius Level

1) There are five cubes in this bag, each cube can be either red OR blue. What are all the possible combinations of colors in the bag?

5 red, 0 blue

1 red, 4 blue

___ red, ___ blue

___ red, ___ blue

___ red, ___ blue

___ red, ___ blue

2) You are asked to draw 5 circles and then some squares. How many squares should you draw if you want a total of 11 shapes?

3) If you combine 5 crackers and 7 crackers, and then eat 10 crackers, how many do you have left?

4) 8 frogs were sitting on a log and 8 more hopped on. The wind blew 3 of them off. How many were left on the log?

5) How many total dots are there on one die?
How many total dots are there on two dice?

Section III

Applying Number Sense

Once children understand the basics of number sense, they are ready to apply that knowledge...

13

Addition and Subtraction

Addition and subtraction? Haven't we been teaching that already?

Yes, you are starting to see the connections! Everything we have done so far will help us add and subtract. Even counting, which is really just adding +1 over and over again.

1...2...3...4... I'm adding!

I know what add/subtract means: more tests where I have to race to answer the most problems in one minute.

No! Timed-tests have a lot of problems. When kids are forced to race against the clock, they don't have time to try new strategies they have just learned.

Timed-tests can be useful for tracking progress from week to week, but they are NOT the best way to teach addition!

5 blue cubes and **1 red cube**

How many when you add them together?

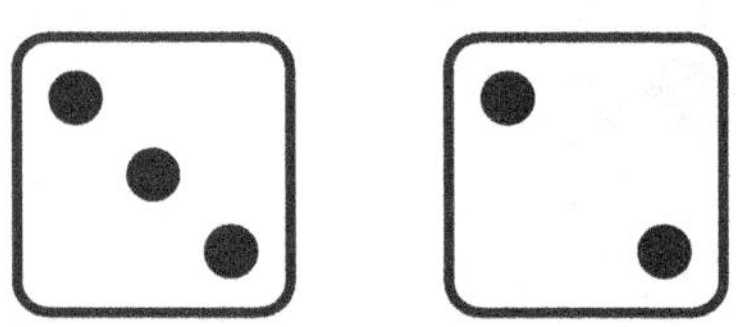

3 and 2, how many altogether?

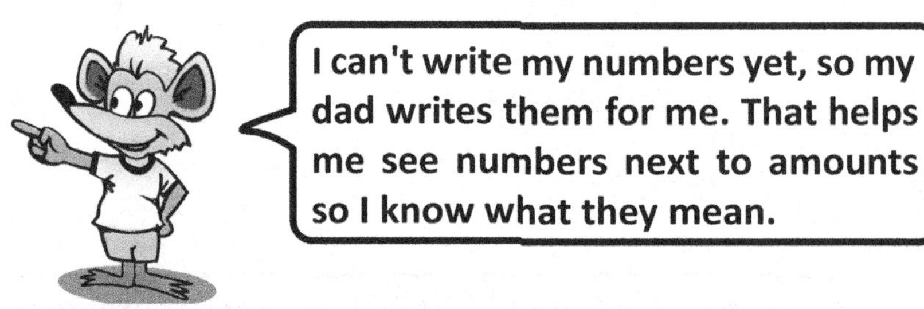

5 bears slept in a cave and 2 more waited outside. How many bears are here?

Abstract Dial Example

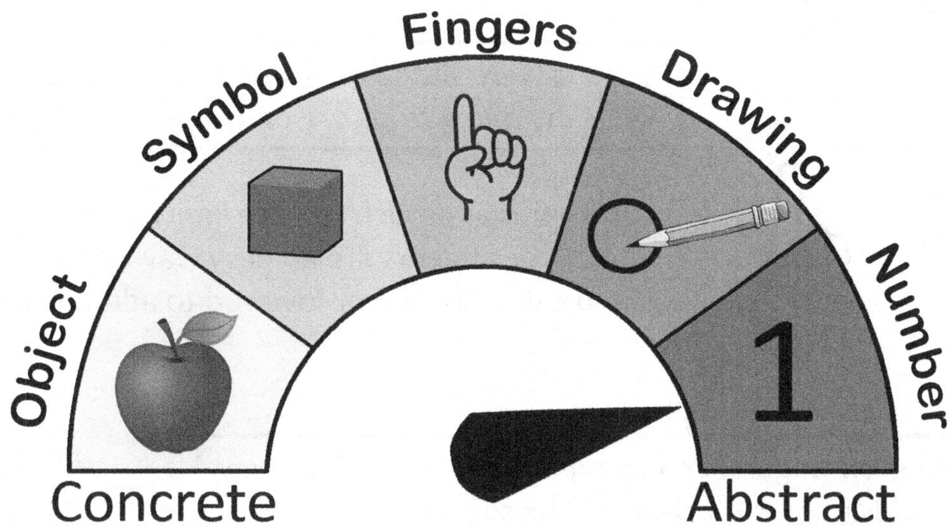

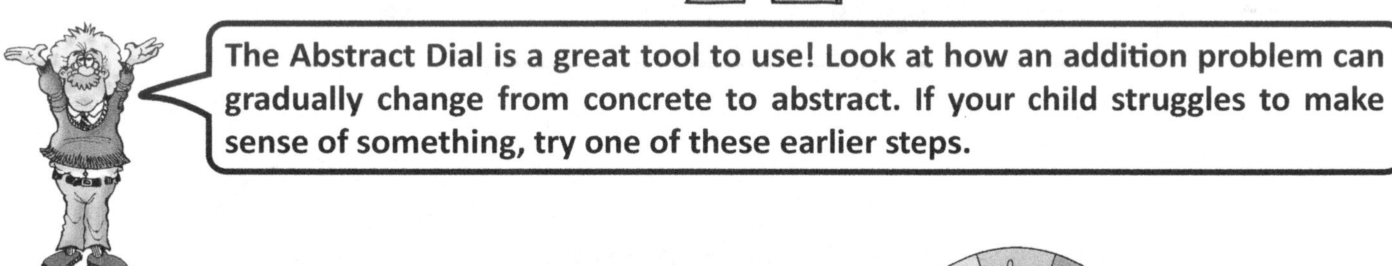

Act it Out

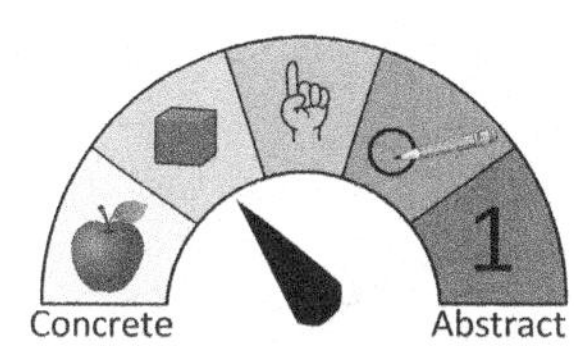

5 frogs were sitting on a leaf. 2 more frogs hopped on. How many frogs are on the leaf now?

Small amounts can be acted out with objects like cubes or crackers. This is the easiest way to understand an idea like addition.

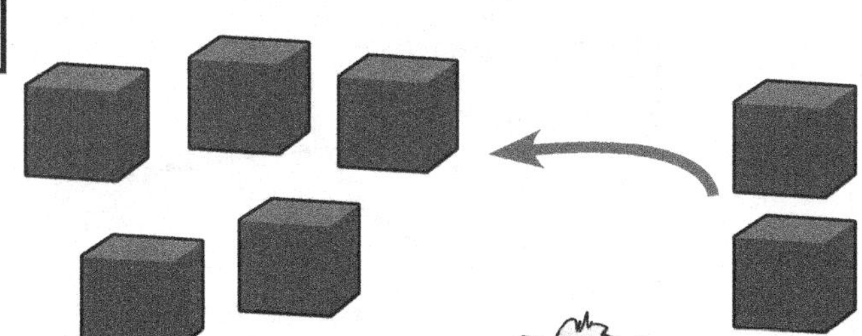

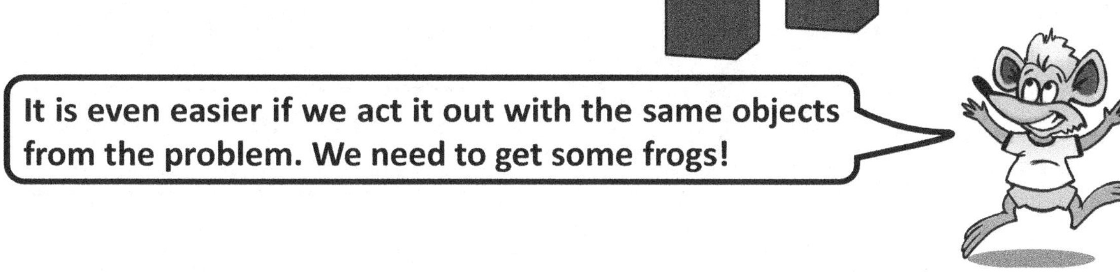

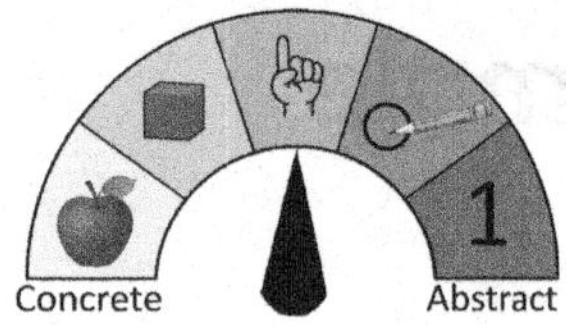

Use Fingers

5 frogs were sitting on a leaf. 2 more frogs hopped on. How many frogs are on the leaf now?

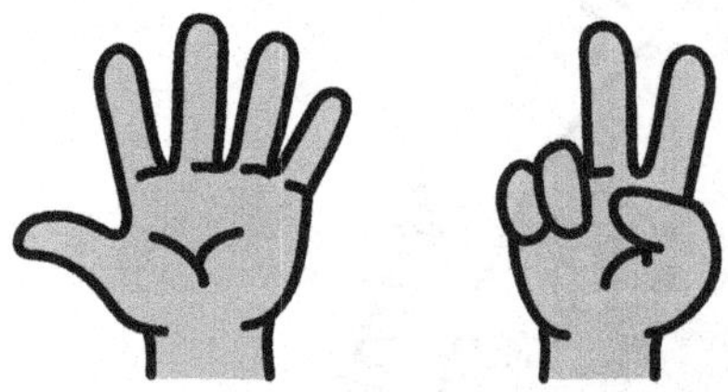

With this approach you are limited to numbers up to 10, but fingers are an easy way to see amounts and can be raised or lowered to add or subtract.

Counting on fingers??? I thought that was a bad way to count! My son's teacher said he shouldn't use them.

When we get older, numbers get bigger and using fingers becomes slow and full of mistakes. But when we are young and still learning about numbers, fingers are a great way to transition from:

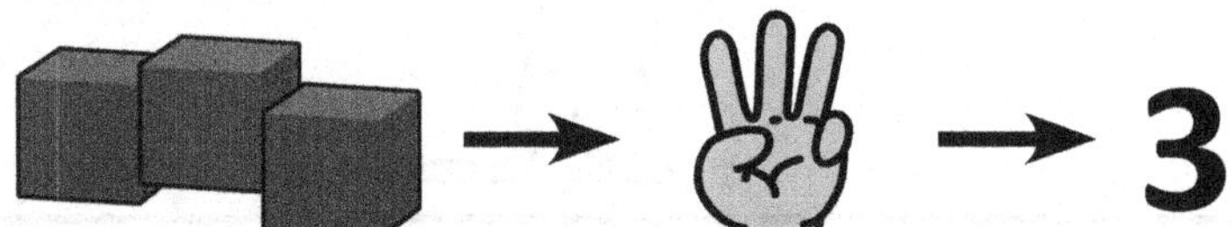

Draw a Picture

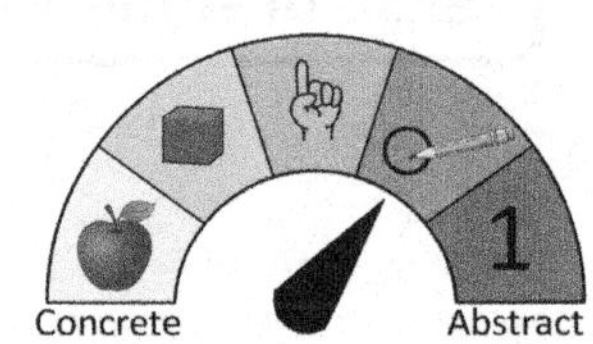

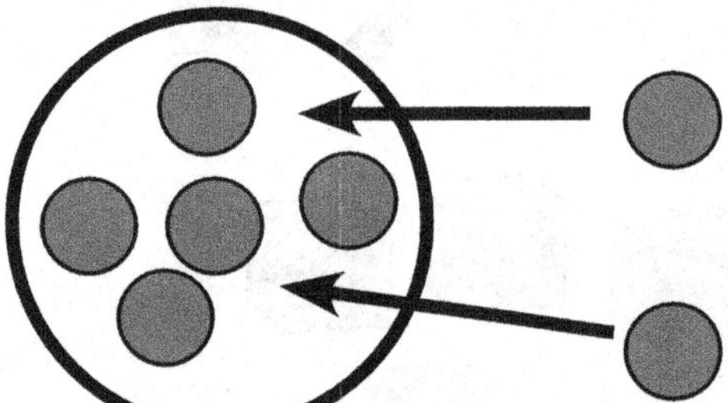

5 frogs were sitting on a leaf. 2 more frogs hopped on. How many frogs are on the leaf now?

Most children can draw basic shapes or tallies. Drawing pictures helps them understand what a problem is asking.

Remember to let me draw the pictures! It makes math more fun.

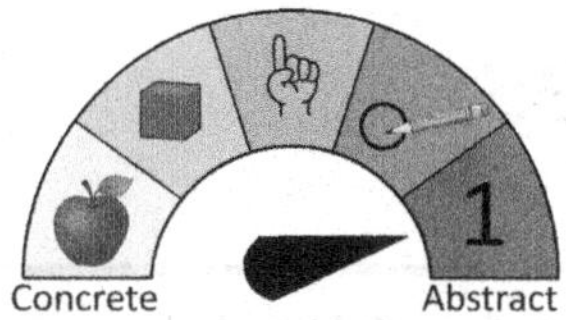

Talk Numbers

5 frogs were sitting on a leaf. 2 more frogs hopped on. How many frogs are on the leaf now?

Talking about numbers (or writing numbers) is the most abstract way to solve a problem.

THIS IS THE GOAL that everyone should aim for. Getting to this abstract level of thinking lets us work with bigger numbers and use advanced ideas like place value.

Adding Strategies

Count Everything (Most Basic)

Just count everything one by one! Most kids start here, but they should be pushed to more advanced strategies as soon as possible.

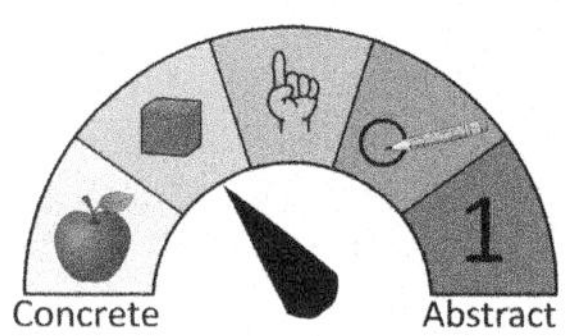

Even when we're still counting by ones, this can be made easier (with objects) or harder (with just numbers).

Start counting from the larger number

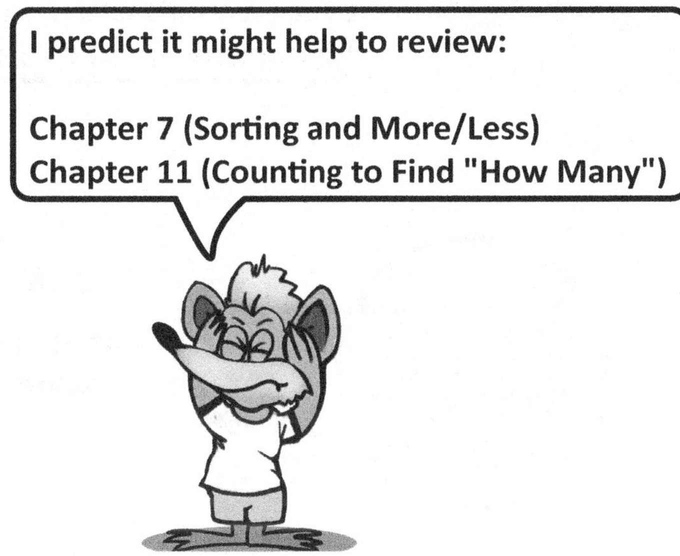

Try starting with both the smaller and larger number. Talk about which one was faster and why it helps to always start counting from the larger number.

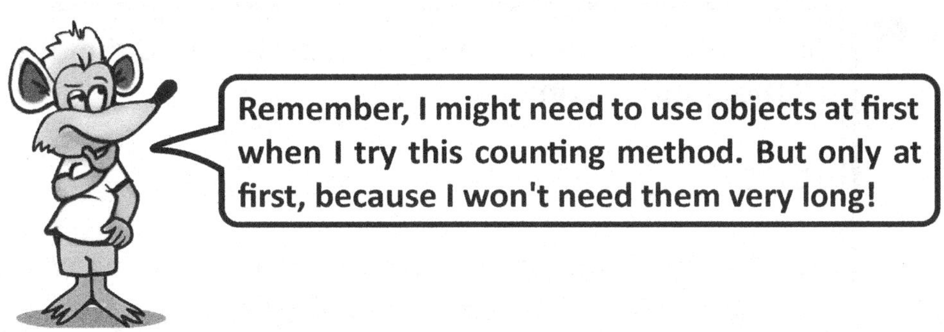

Subtraction

Oh no, I forgot about subtraction! I hope that isn't very hard to learn because I don't know if my brain is big enough.

Luckily, subtraction is just like doing addition—only backwards! A lot of the strategies will be similar.

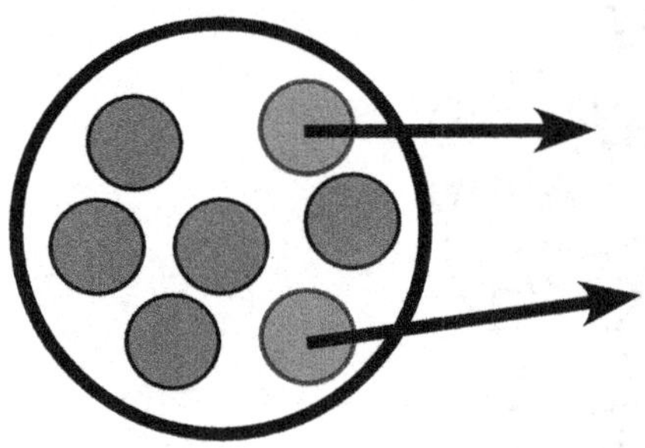

Take Away

Start at one number and count backwards to "take away".

Compare

Common for "how many more" and "how many less" problems. Looking at the distance between numbers is another way to think about subtraction.

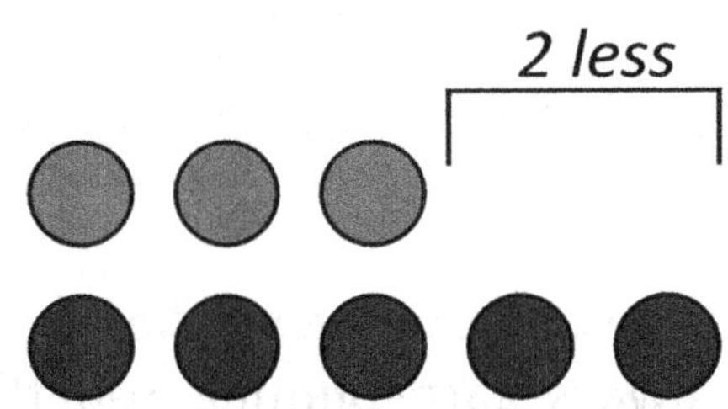

Most kids are used to one type of problem, so that is usually all they improve on:

15 + 9 = ____

But it is VERY important to give them problems where the "unknown" is in different places, like this:

15 + ____ = 24   ____ + 9 = 24   24 = ____ + 9

This helps them become more flexible with numbers, which gets them ready to adapt to any type of problem.

Addition and Subtraction

Level 1

1) There is 1 banana on the counter and we picked up 2 more at the grocery store. Will we have enough for me and each of my brothers?

2) My cat Skinner has 6 toys, then my parents brought over 5 more. How many toys does my cat have now?

3) I'm starting a penny collection! On Monday I found 3 pennies, on Tuesday I found 5 pennies, and on Wednesday I found 2 pennies. How many pennies did I collect this week?

4) I had 10 gummy bears and ate some of them until there were 7 left. How many did I eat?

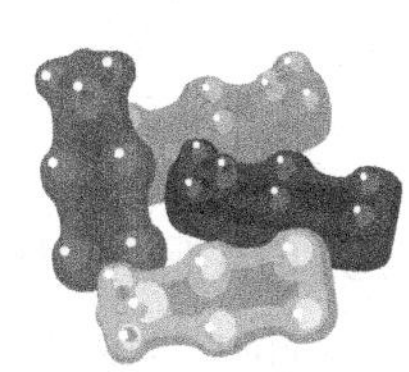

5) My rock collection has 10 beautiful rocks. I gave 7 of them to my sister for her birthday. How many rocks do I have left?

6) I have 12 pennies in a box. I gave 2 of the pennies to my dad and then gave another one to my mom. How many pennies are in the box now?

7) I brought 12 blue M&Ms to lunch and my friend brought 10 red M&Ms. If we combine them, how many pieces of candy will we have?

8) I took 26 steps and then I took 5 more steps. How many steps have I taken altogether?

9) Jasmine ate 3 of her 24 grapes. How many grapes does she have left?

10) On Christmas Day, I opened 4 presents and there were 16 left. How many presents were under the tree before I started opening them?

Addition and Subtraction

Level 2

1) A mouse found 20 raisins and ate 6. He then gave 7 to his sister. How many raisins are left?

2) 8 apples can be split into two groups in many different ways. One way is a group of 6 and 2. Draw 2 other ways to split up 8 apples.

3) A rat climbed up 10 steps on a ladder, then climbed down 7 steps. Then he went up 2 steps. Where is the rat on the ladder now?

4) Jesse wanted to collect 27 black rocks. He found 20 so far. How many more rocks does he need?

5) Rachel and Lauren need to collect 45 pinecones for a school project. Rachel has 15 and Lauren has 20. How many more do they need?

6) A full deck of cards has 52 cards in it. Kevin's deck has only 42 cards. How many cards are missing?

7) Pam drew 25 circles on a piece of paper. Then she drew an ant in 15 of the circles. How many circles are left with no ant in them?

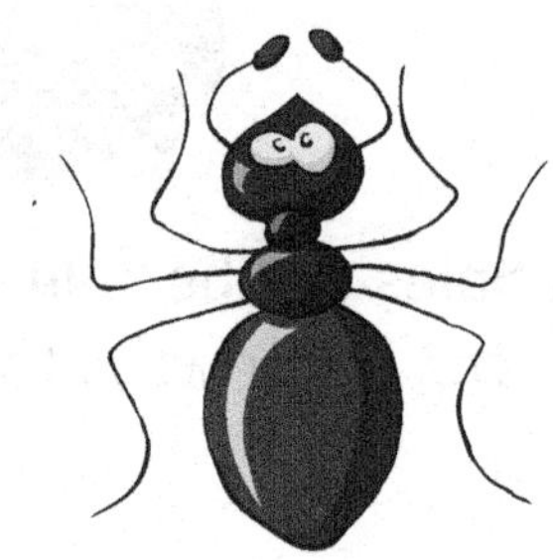

8) Adelyn has 20 pieces of candy in her Halloween bag. She ate 5 pieces, gave away 5 pieces, and then bought 3 more at the grocery store. How many pieces of candy are in the bag now?

9) I took 26 steps, then I took 10 more steps. Then I rested for a while before taking 4 more steps. How many total steps have I taken?

10) 16 apples were split into two groups. If one of the groups had 7 apples, how many apples were in the other group?

Addition and Subtraction

Level 3

1) Warren has a chessboard with 64 squares. He put a piece of candy on each square until he ran out of candy. There were 18 squares left. How many pieces of candy did he start with?

2) A chef needs to cook dinner for 12 people. At the last minute, the plans changed and now he needs to cook for 35 *more* people. How many meals does he need to prepare?

3) A box had 30 marbles inside it. Dave gave 10 marbles to a friend and then lost 15 marbles. He bought 7 more with his allowance and put them in the box. How many marbles are in the box now?

4) Pam set the table by putting out 32 plates. She started putting napkins on some of the plates until there were 12 plates left without napkins. How many napkins did she put on the plates?

5) Billie and her sister currently have a total of 13 cookies. If they **each** want 9 cookies, how many more do they need to buy?

6) A deck of cards is supposed to have 52 cards in it. How many cards are missing if only 12 cards are in the deck?

7) A chess set has 16 pawns, 2 kings, 2 queens, 4 bishops, 4 knights and 4 rooks. How many total pieces are in a chess set?

8) Michael had a pile of 16 cheese wedges. He ate 8 of them and then ate 4 more. How many cheese wedges does Michael have now?

9) I picked up 40 shells at the beach and then threw 10 in the water. I picked up 15 more and then threw 5 in the water. How many shells do I have now?

10) Morgan and her 2 brothers have a total of 15 tulips. If they **each** want to plant 7 tulips, how many more do they need to buy?

Addition and Subtraction

Genius Level

1) Warren has a chessboard with 64 squares. He put a piece of candy on 17 squares and then put a penny on 18 different squares. How many squares do not have any candy or money on them?

2) What is 20 + 23 ?

3) ____ - 28 = 40 ?

4) Alice is in a room with 5 spiders, 4 grasshoppers, 3 dogs and 2 ducks. How many legs are there in the room? (Don't forget Alice's legs!)

5) 4 people have a total of 33 checkers. If they each need 12 to play the game, how many more do they need to buy?

6) My trading card deck should have 112 cards in it, but after I counted them I realized I need to look for 37 missing cards the next time I clean my room. How many cards did I count in the deck?

7) A box had 1235 puzzle pieces inside it. Another box holds 2,421 pieces and a third box holds 1004 pieces. How many puzzle pieces would I have if I dumped all the boxes together?

8) What do you get if you add the first 10 numbers together (starting with 1)?

9) I made $40 working as a bellhop but spent $6 on lunch that day. On the way home I found $16 on the sidewalk, but left $10 there for the next person to find. How much money did I end up with today?

10) Morgan and her 4 brothers have a total of 15 books on their bookshelf. They each want to have 12 books, so how many more should they buy at the book fair?

14

Place Value

When is Someone Ready?

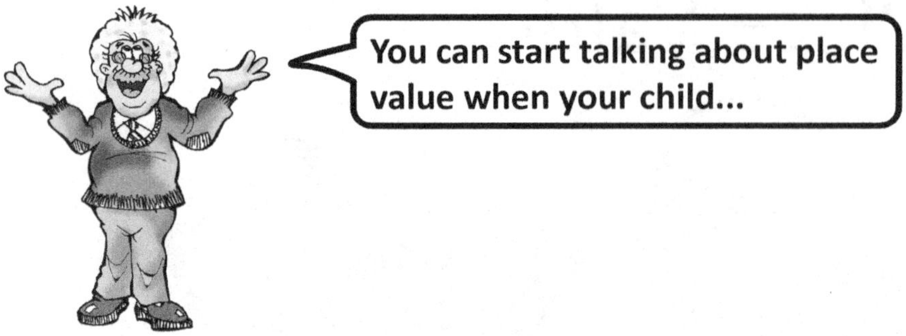

Sees groups of ten as "countable"

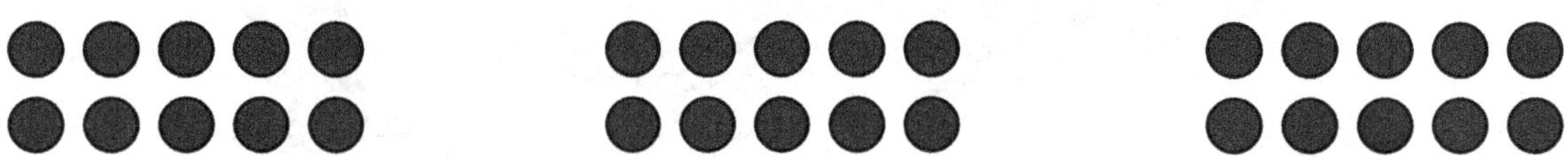

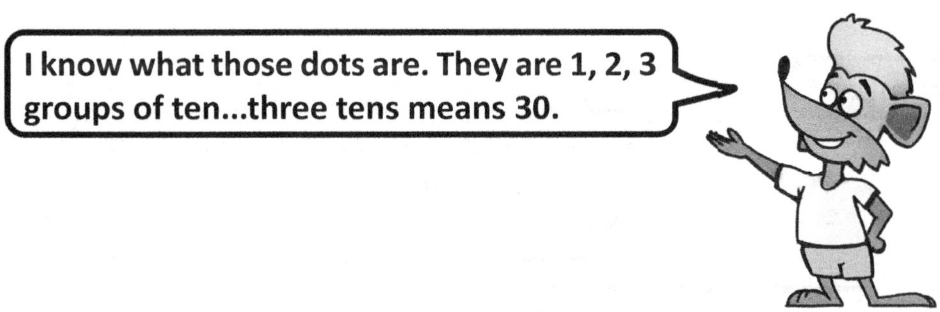

Understands that "17" is made up of "10 and 7 more"

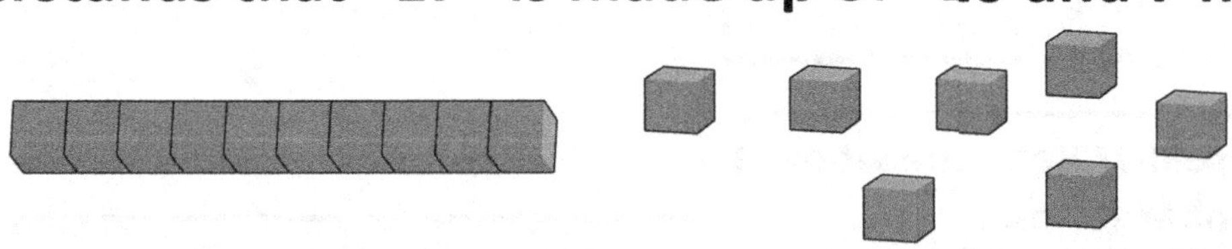

Can skip count by 10s and 100s

Can add "tens" and "ones" separately

54 + 32

Warning signs that your child is NOT ready to talk about Place Value:

- **Cannot count without making mistakes.**
- **Doesn't understand that "6" can be built from 4 & 2 (or 5 & 1, etc).**
- **Doesn't realize when a problem is asking for addition or subtraction.**

If your child does not have these basic skills, it will be difficult to teach more complicated ones. She might be able to answer some of your place value questions, but chances are she won't really understand them.

Students first learn things like 17 is really "10 and 7 more." This leads to expanded form (342 = 300 + 40 + 2) and eventually fractions, decimals, and percents!

But why is understanding Place Value so important if I can just memorize a list of steps to find an answer?

Knowing how to break apart numbers into their Place Value parts lets us work quickly and flexibly with numbers!

Place Value means that the value of a number changes depending on its location, or place.

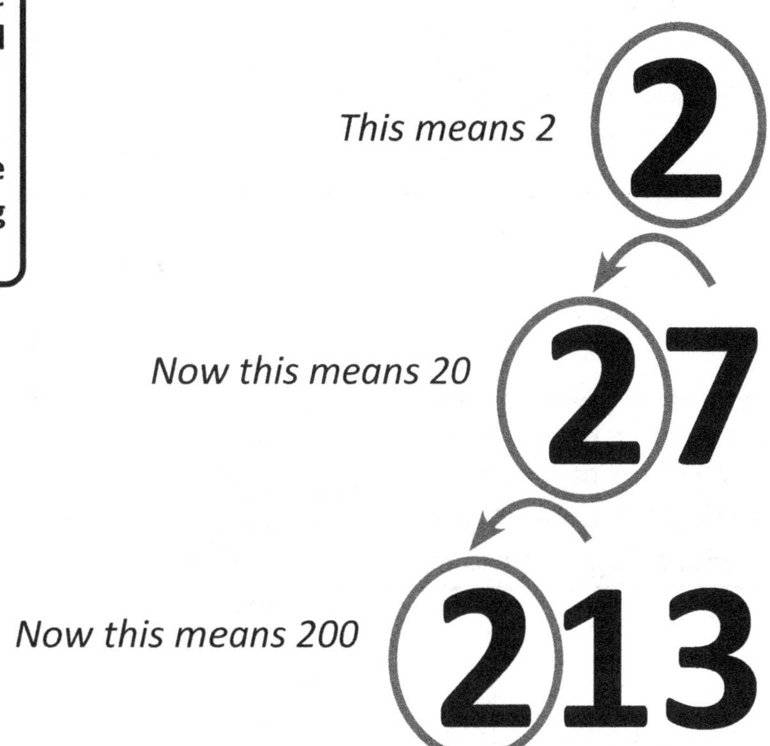

Place Value Tools

Place Value Blocks

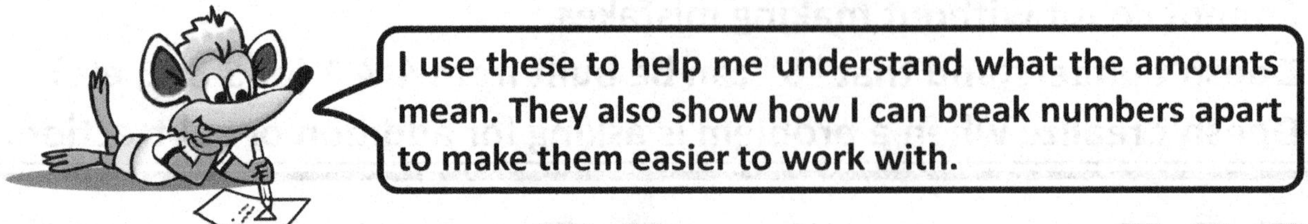

27 + 33 →

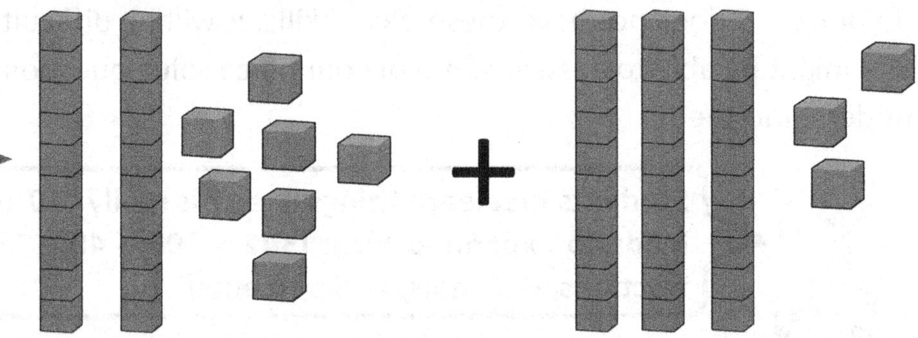

Words matter! Some people call place value blocks things like **"skinnies and bits"** or **"longs and flats."** But those words don't mean anything math-related, so every time a brain hears *"skinny,"* it has to translate it to *"ten."* Use a name that matches the amount so the word instantly communicates the right number.

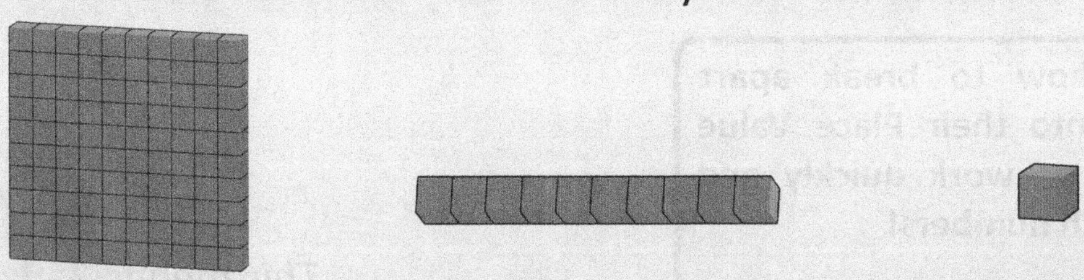

Hundred block *Ten Strip* *One block (or cube)*

Place Value Cards

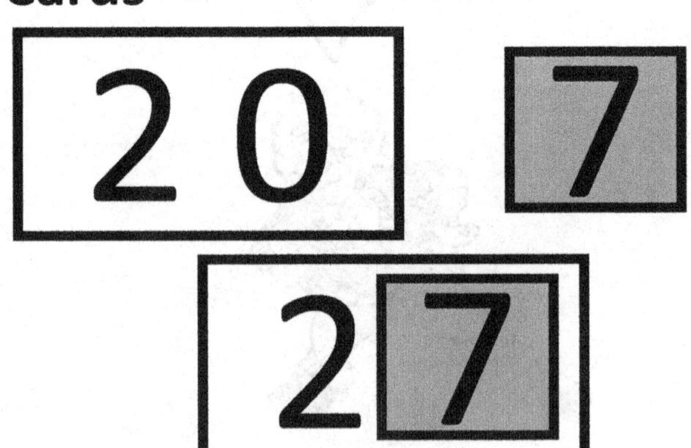

(Copy and print your own on page 335!)

Place Value Mats

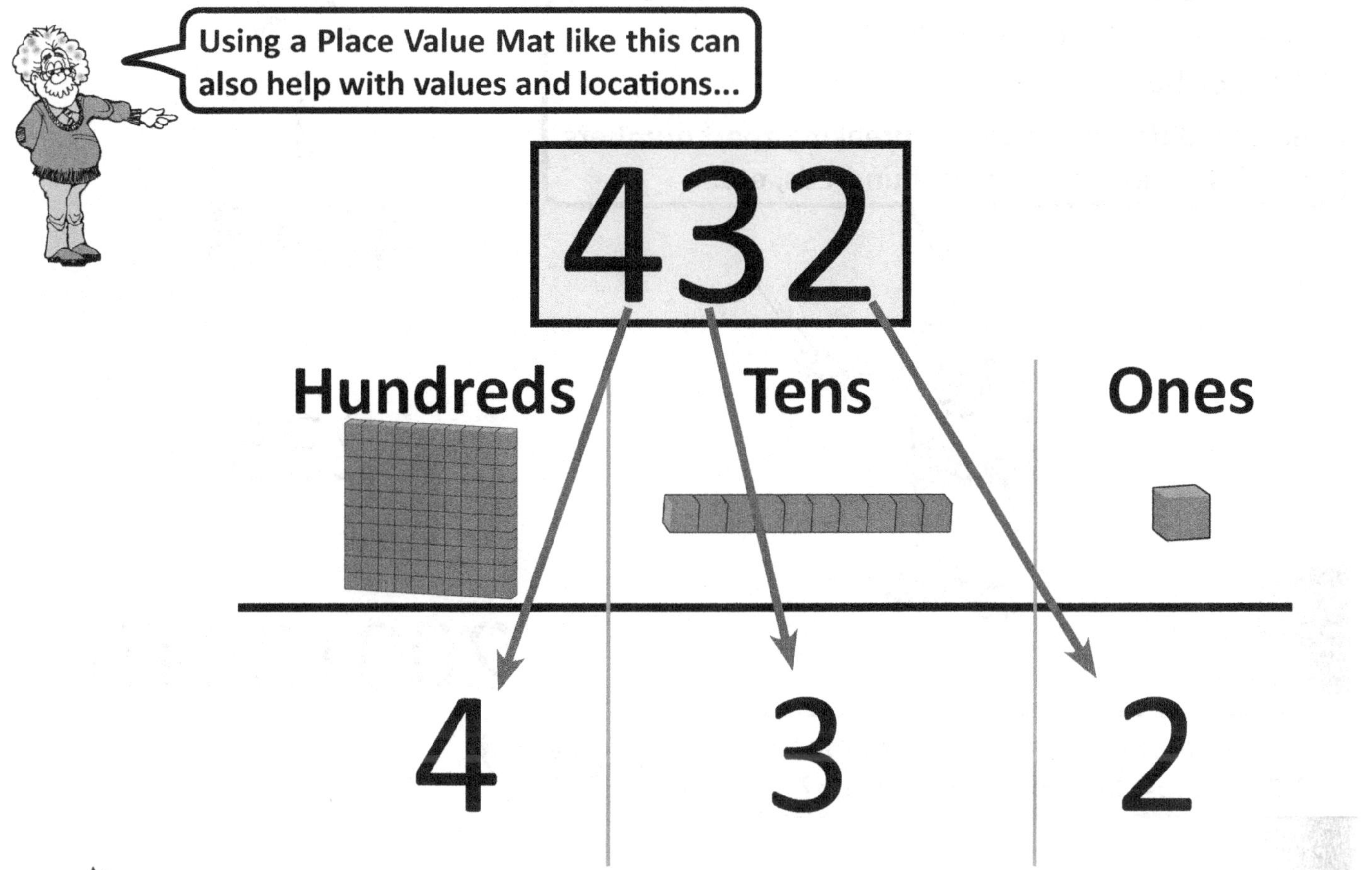

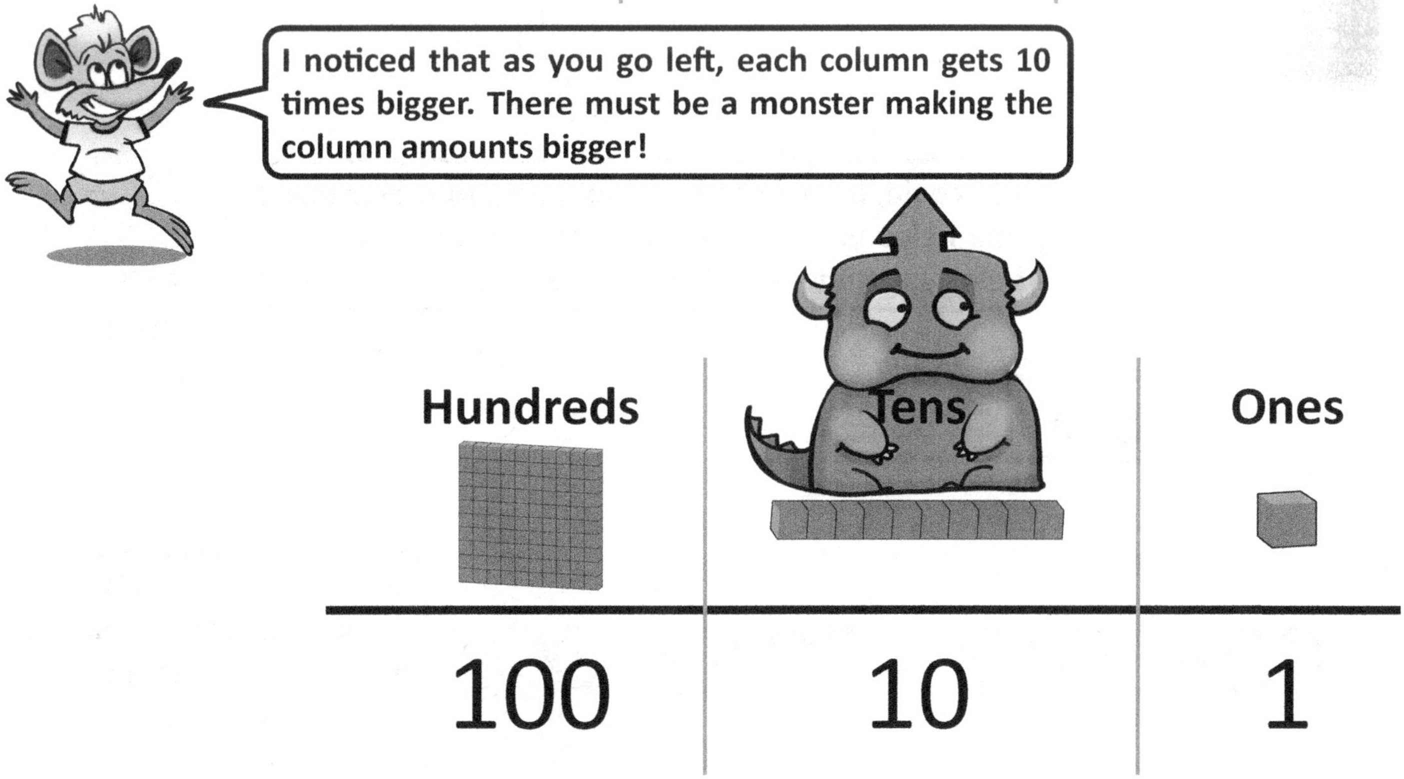

(See page 337 for a blank Place Value Mat.)

Expanded Form

Expanded form is just like the idea:

"7 can be built from 5 & 2"

The only difference is we are breaking apart numbers in specific ways: ones, tens, hundreds, etc.

43

↓

40+3

235

↓

200+30+5

Why would I need to break numbers up? Can't I just line them up and add?

You could, but sometimes expanding numbers like this helps you work quicker and smarter. It also helps you understand what large numbers mean.

<u>237</u>+<u>72</u>

↙ ↘

200+30+7 70+2

30 and 70 make 100. That means we have 3 hundreds, a 7, and a 2. The answer is 309, I didn't even need to write it down!

Place Value

Level 1

Use place value concepts like "separating the tens and ones" or "expanded form" to solve the following problems...

1) Lucas is 10 years old. In 3 years, Lucas will be ____ years old.

2) If Liam had 32 marbles and found 20 more, how many marbles would he have?

3) Noah's mom asked him to pick up 75 pinecones from the ground. If Noah picked up 55 pinecones, how many more does he need to pick up?

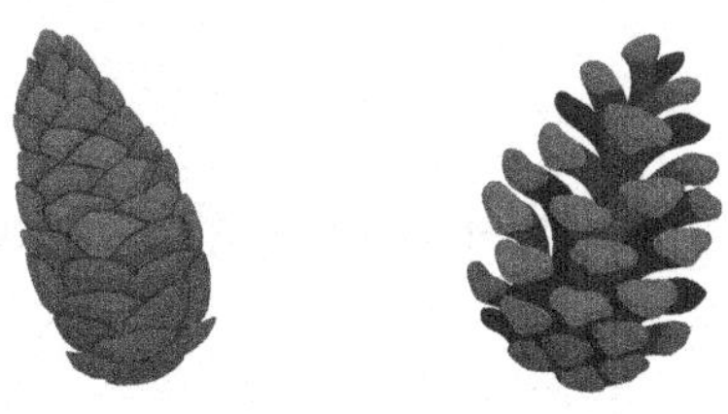

4) 400 + 40 + 4 = ____?

5) 562 - 40 = ____

210 - 10 = ____

6) Ava counted 82 cows on one side of a field and then counted 17 cows on the other side. How many total cows are there?

7) 80 can be written as 40 + 40 or 50 + 30. What are two other ways that 80 can be written?

8) In the number 890, what value does the number "9" stand for?

9) Together, Oliver and his dog weigh 104 pounds. If Oliver weighs 52 pounds, how much does his dog weight?

10) If a carpenter has 249 nails, it means that she has

a pile of 200 nails +

a pile of ____ nails +

a pile of ____ nails

Place Value

Level 2

Use place value concepts like "separating the tens and ones" or "expanded form" to solve the following problems...

1) Change the following into one number: 5000 + 300 + 20 + 1

2) Emma has 68 dollars in her savings account. How much will she have left if she spends 50 dollars?

3) In the number 8022, what value does the number "8" stand for?

4) Tifanie's piggy bank had 68 cents in it, but she took out 15 cents to throw into a wishing well. How many cents are left in her piggy bank?

5) A guitar cost $108 and a violin cost $308. How much money will I need if I want to buy both?

6) If a carpenter has 8292 nails, it means that he has

a pile of 8000 nails +

a pile of ____ nails +

a pile of ____ nails +

a pile of ____ nails

7) Change the following into one number: 50,000 + 3000 + 700 + 10 + 1

8) 890 + ____ = 1000

3452 + 300 = ____

9) 7260 = 7 thousands, ____ hundreds, ____ tens, and 0 ones

10) Change the following into one number: 90,000 + 9000 + 900 + 90 + 3

Place Value

Level 3

Use place value concepts like "separating the tens and ones" or "expanded form" to solve the following problems...

1) Isabella walked 125 steps and then walked another 524 steps. How many steps did Isabella walk altogether?

2) 64 can be written as 68 minus 4. What are two other ways to write 64 using subtraction?

3) Sophia had a goal to walk 900 steps in one hour. If she walked 325 steps, how many more must she walk to reach her goal?

4) ____ - $430 = $470?

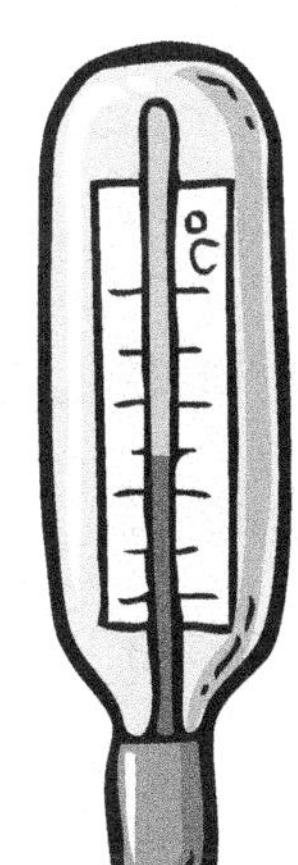

5) The temperature was 102 degrees, but then it dropped 30 degrees during the night. What is the new temperature?

6) If Samuel has 8,642 pennies in his coin collection, it means he has:

a pile of ____ pennies +

a pile of ____ pennies +

a pile of ____ pennies +

a pile of ____ pennies

7) In expanded form, the number 35,625 is equal to: 30,000 + 5000 + 600 + 20 + 5

What is the number 88,888 written in expanded form?

8) What is the number 3,450,000 written in expanded form?

9) 4580 + 6000 = ____

3490 + ____ = 9490

10) If Orin has 125,100 pennies in his coin collection, it means Orin has...

a pile of 100,000 pennies +

a pile of ____ pennies +

a pile of ____ pennies +

a pile of ____ pennies

Place Value

Genius Level

Use place value concepts like "separating the tens and ones" or "expanded form" to solve the following problems...

1) Jacob had a goal to walk 10,000 steps. If he has walked 6,520 steps, how many more must he walk?

2) 3 tens and 7 ones = 2 tens and ____ ones?

3) Complete the blanks:

   562 = ____ hundreds + ____ tens + 72

4) The boiling point of water is 212 degrees and the freezing point of water is 32 degrees. How much higher is the boiling point of water than the freezing point of water?

5) The Mariana Trench is in the Pacific Ocean and is 36,070 feet deep. Mt. Everest is 29,030 feet tall. How much distance is there between the top of Mt. Everest and the bottom of the Mariana Trench?

15

Multiplication and Division

Remember from the Place Value chapter how this "2" really means "20"?

When multiplying, some numbers have hidden meanings too. That is why it can be tricky to learn at first!

I know this one! That "3" really means "3 GROUPS OF..."

That's right! Kids will learn this idea much better when they can first visualize what it means to "multiply." This can be done with drawings or objects, depending on the age and skills of the child.

The key idea to remember for Multiplication (and Division) is "equal groups." Look for when these groups happen in real life and call them out as examples of Multiplication.

For example... socks, sticks of gum, coins

Start with Pairs

Pairs are the easiest place to start because we are all familiar with things like socks and gloves.

Connect that with the idea of multiplication!

3 x 2

↓

3 (groups of) x 2 (socks)

= 6

What if I wanted to flip the numbers and make 2 groups with 3 socks each? Hey, the total number of socks doesn't change! That must be important...

Arrays

3 groups of 2

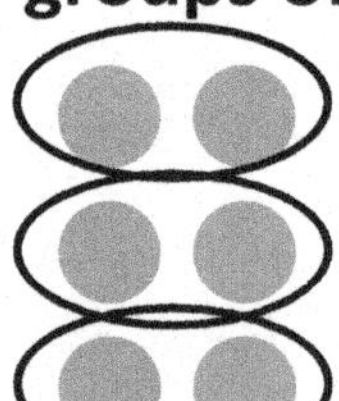

We can also use an Array to show what multiplication means. These can be rotated to help show that the total amount stays the same, even if the numbers flip (just like with Addition).

2 groups of 3

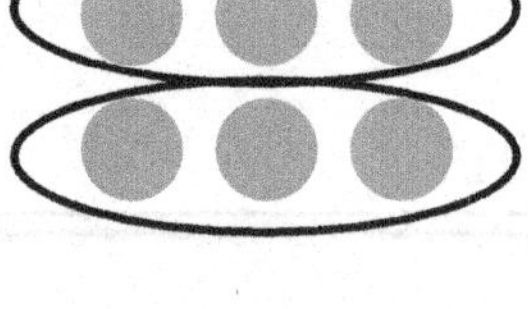

Arrays can also show how Multiplication is similar to "Skip Counting."
2...
4...
6...
8...
10...
12...
2 x 6

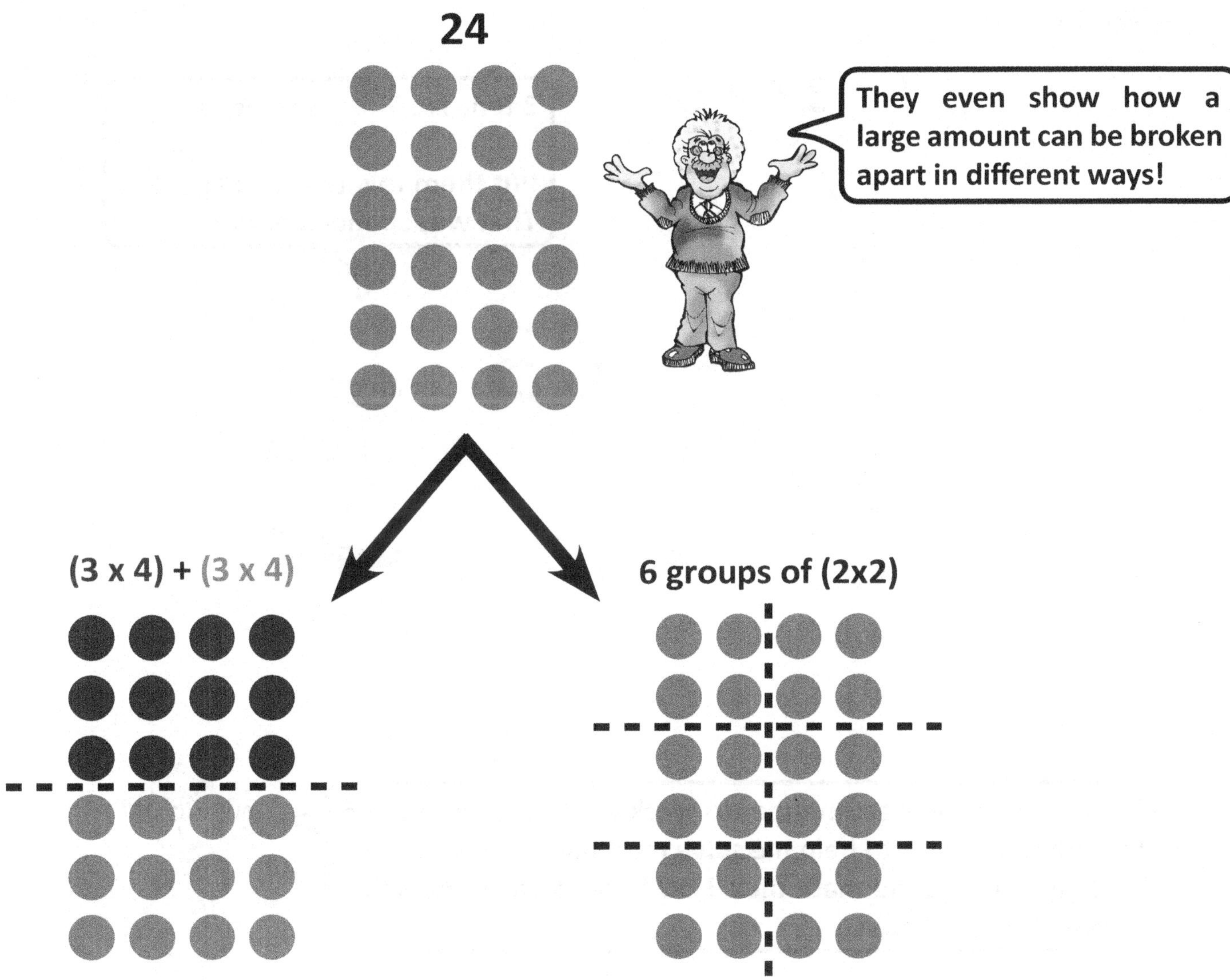
24
They even show how a large amount can be broken apart in different ways!
(3 x 4) + (3 x 4)
6 groups of (2x2)

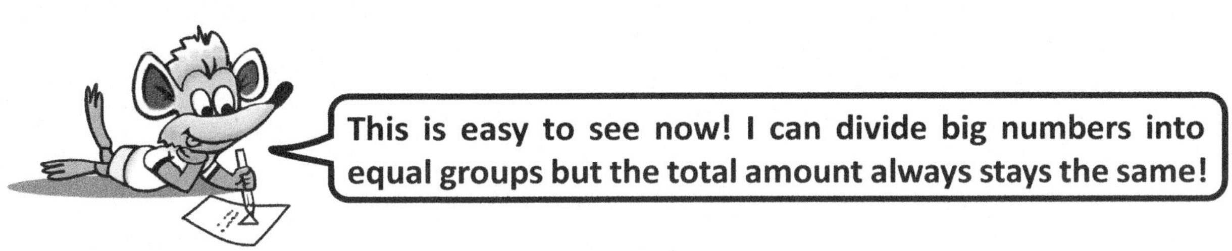
This is easy to see now! I can divide big numbers into equal groups but the total amount always stays the same!

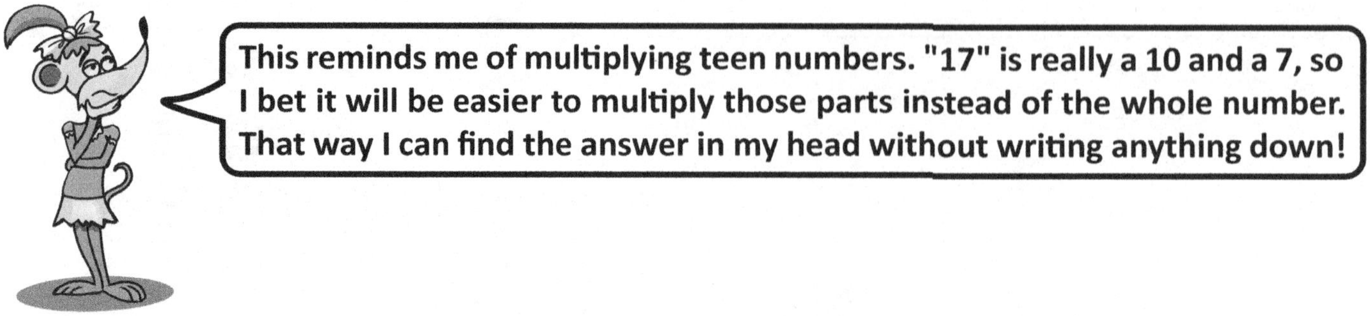
This reminds me of multiplying teen numbers. "17" is really a 10 and a 7, so I bet it will be easier to multiply those parts instead of the whole number. That way I can find the answer in my head without writing anything down!

17 x 3
x 3
3 tens is 30
3 x 7 is 21
Put them together and I get 51.
That was so much easier!

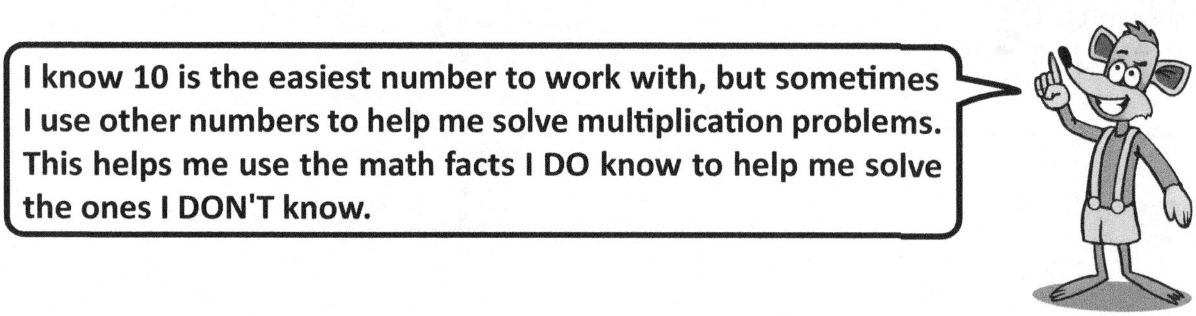
I know 10 is the easiest number to work with, but sometimes I use other numbers to help me solve multiplication problems. This helps me use the math facts I DO know to help me solve the ones I DON'T know.

Division

Division is the opposite of Multiplication (just like Subtraction is the opposite of Addition).

Division means that instead of finding the total amount, we are figuring out "how many groups" OR "how many per group."

I'm guessing you are going to show me an array again...

Not yet! Kids can understand Division by first learning how to share an amount equally. Being "fair" gets to the important idea of "equal groups."

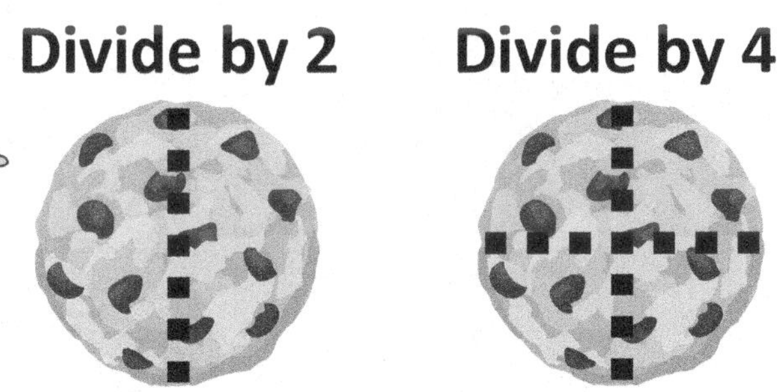

This can also happen with a set of objects! Have the child pass them out one at a time, then count each pile to see how they are the same.

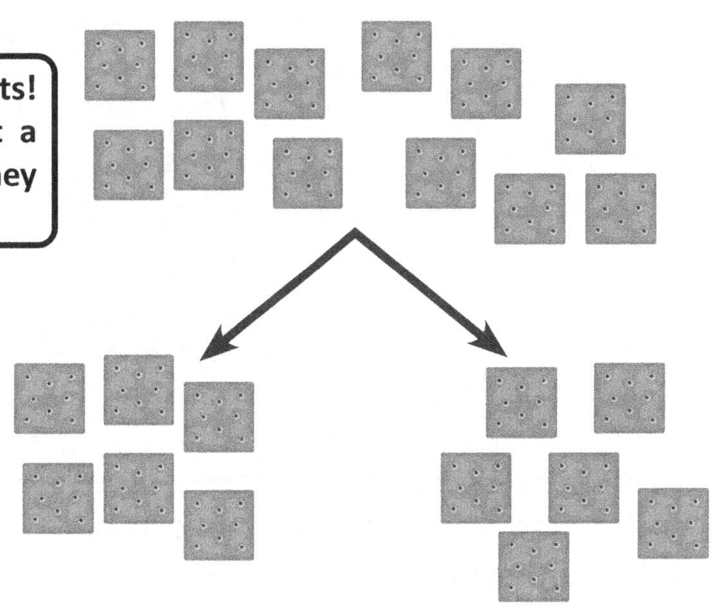

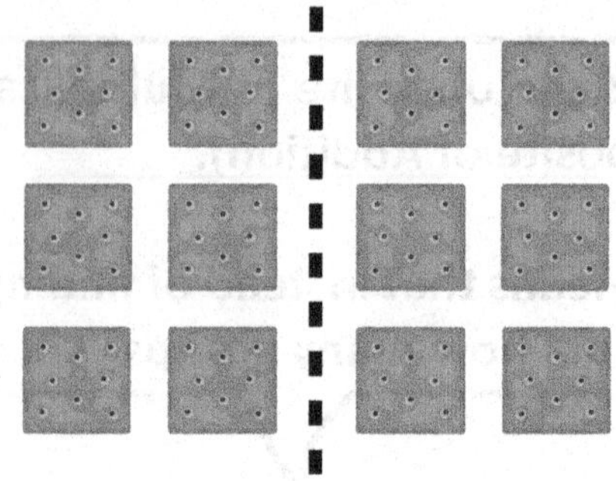

24 ÷ 4 = 6

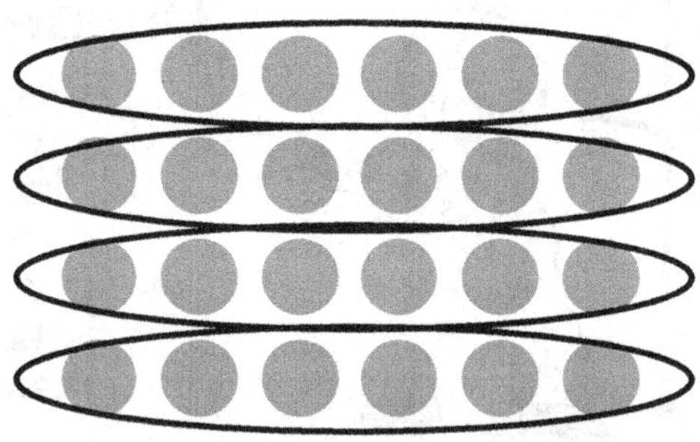

Divide into 4 groups
(6 in each group)

24 ÷ 6 = 4

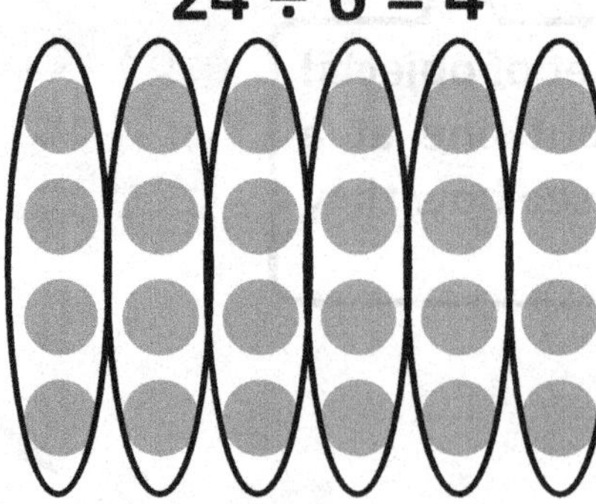

Divide into 6 groups
(4 in each group)

> Drawing a picture like this can help when we get stuck on a problem, but it shouldn't be our approach every time.

Shapes as Arrays

Keep in mind that regular shapes can work like arrays once a child understands the basic idea of Multiplication and Division. These are much quicker to draw and work with!

I'm using shapes to help solve my homework until I understand multiplication a little better. Here are some examples...

5 x 6

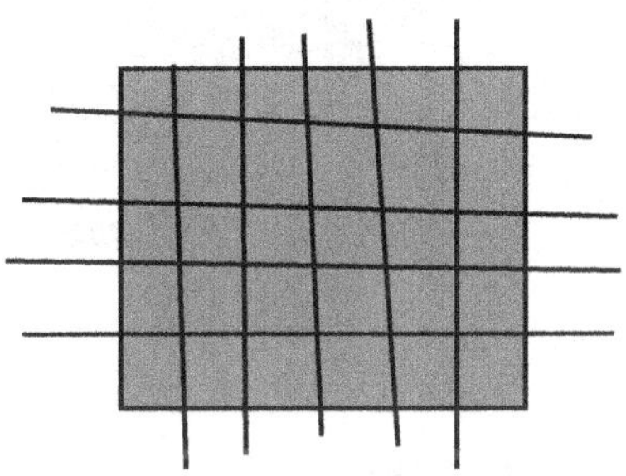

I'll make a grid and count the squares for 30! (this takes too long)

5 x 6

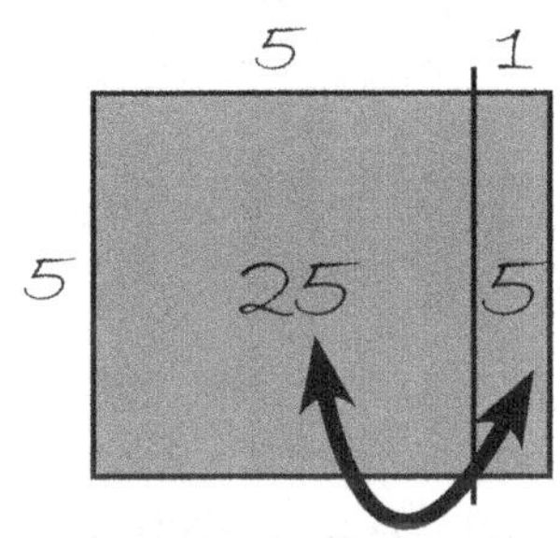

I'll split the 6 into "5 and 1" and add the parts.

5 x 5 = 25
5 x 1 = 5   *30!*

17 x 3

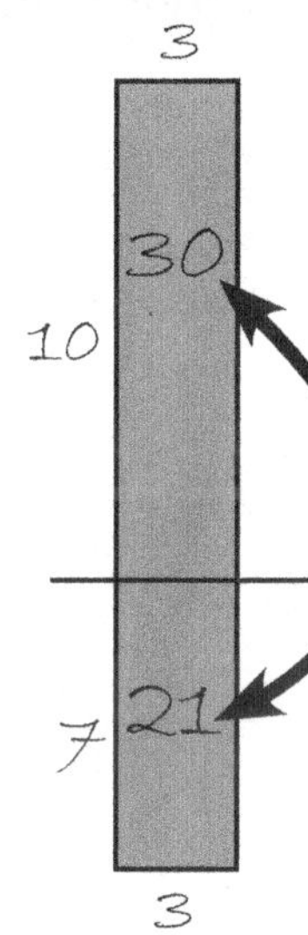

I'll split up the 17 and add the parts.

10 x 3 = 30
7 x 3 = 21

51!

24 ÷ 6

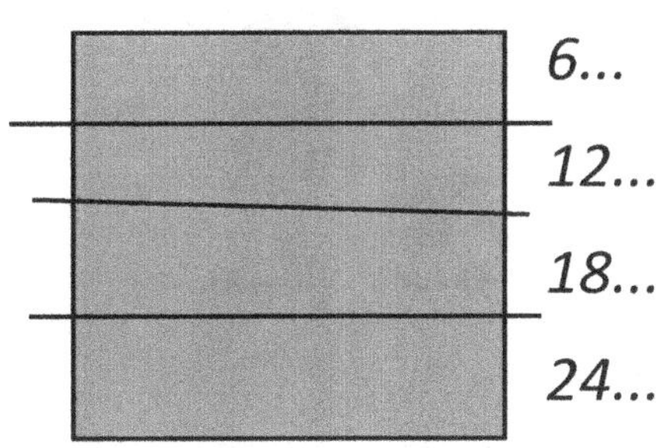

I'll skip count by 6 and mark the groups. It took FOUR groups to get to 24, so the answer is 4!

The best way to understand Division is to first understand *Multiplication*. This will help us become automatic with our math facts and see even more connections between numbers.

Multiplication and Division

Level 1

1) How many legs do 4 cats have?

2) I have 3 boxes and they each have 8 toys in them. How can I find the total number of toys I have?

3) Solve 13 x 4 by breaking apart the 13. Make a drawing to show what you did.

4) Soccer teams usually have 11 players each. How many shoes would a soccer team need?

5) How would you share these cookies fairly with your brother?

6) What two multiplication facts could this show?

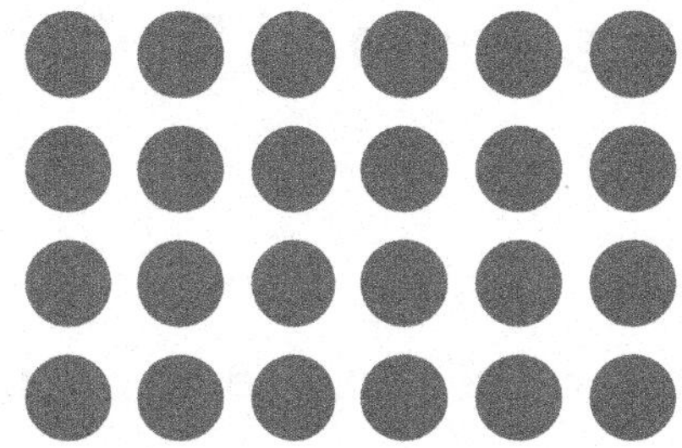

7) Draw an example of 7 x 3.

8) What multiplication fact does this show?

9) How many total arms do 3 greedy fleas have?

10) What multiplication fact does this show?

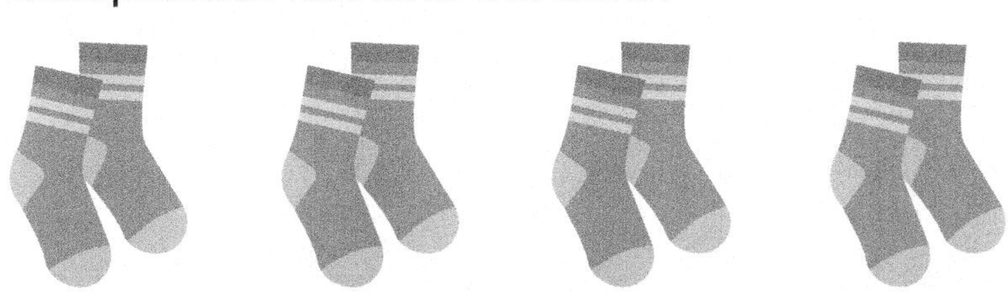

Multiplication and Division

Level 2

1) How can you use **4 x 8** to help you find the answer to **8 x 4?**

2) If 17 x 64 = 1088, two things must also be true:

1088 ÷ ___ = ___   **1088 ÷ ___ = ___**

3) Kelly counted 8 cookies on her plate, multiplied this by 3 plates, and said "there are 24 cookies here." Is she right? Why or why not?

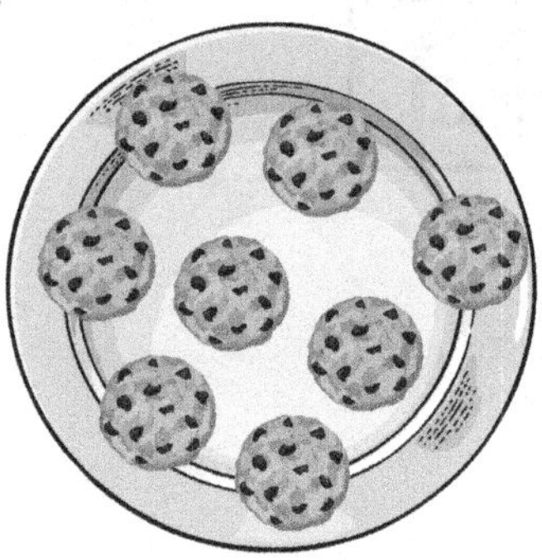

Kelly's Plate

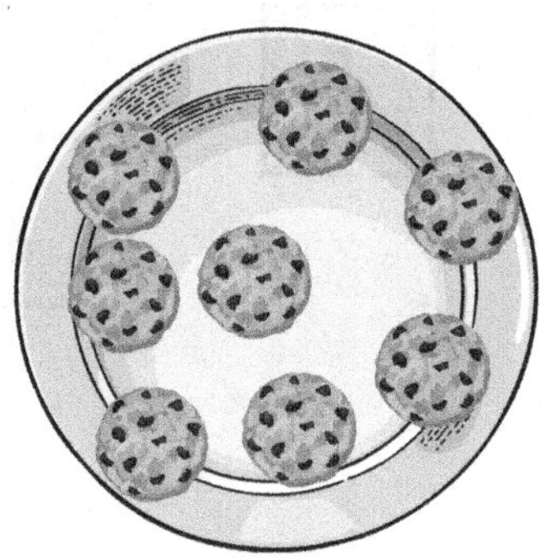

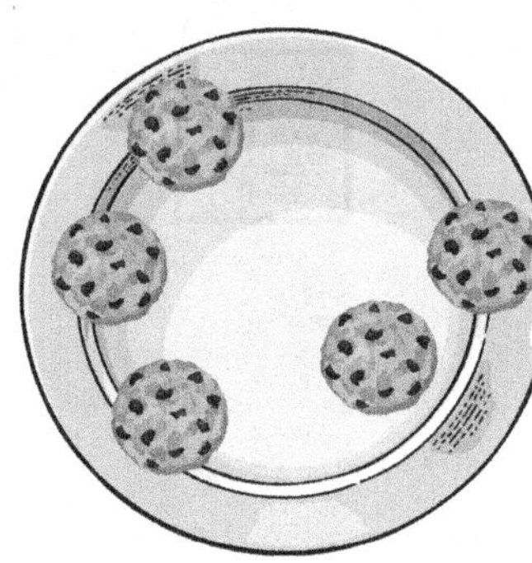

4) How you would share these mini-pizzas fairly between 3 people?

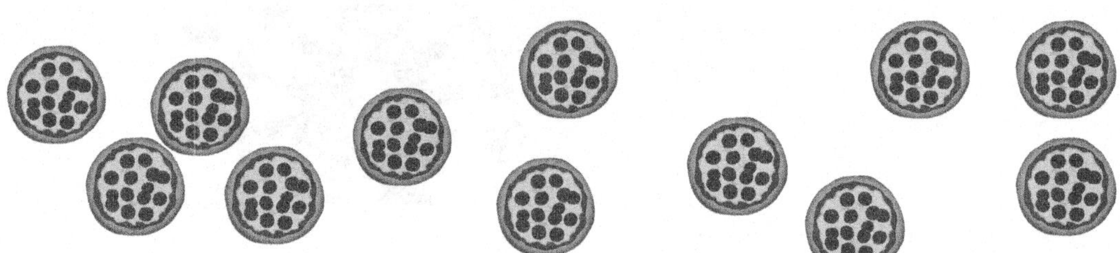

5) A farmer grew a total of 40 carrots in rows of 8 each. How many rows of carrots did he have?

6) What two multiplication facts could this show?

7) Eggs usually come in cartons of 12 (called a dozen). How many cartons do I need to hold 48 eggs?

8) A chessboard has 8 squares in a row and there are 8 rows. How many squares are on a chessboard?

9) A chess set has 16 pawns, 2 kings, 2 queens, 4 bishops, 4 knights and 4 rooks. How many bishops are in 6 chess sets?

10) How many socks do 55 people need?

Multiplication and Division

Level 3

1) A paper chessboard with 64 squares is being cut up into groups of 4-squares each. How many of these groups will there be?

2) My house has 3 swarms of ladybugs with 34 in each group. How many total ladybugs are there?

3) If 3 people each had 17 marbles, how many marbles are there all together?

4) Brianna drew 60 pictures of her favorite cat. She gave all the drawings away to her 4 friends. How many drawings did each friend receive?

5) Lenny's bedroom is in the shape of a rectangle. The floor has 7 tiles along one side and 6 tiles along the other side. How many tiles are there on Lenny's floor?

6) How could you use this array to show 48 ÷ 8?

7) Bananas cost 35 cents per pound. If I bought *only* bananas and got charged $1.40, how many pounds did I buy?

8) If a cook had 5 dozen eggs and used 5 eggs each day, how many days would the eggs last?

9) 1,000,003 + 1,000,003 + 1,000,003 = ?

10) Each person has 8 fingers, 2 thumbs and 10 toes.
In a group of 11 people, there are:

____ fingers, ____ thumbs and ____ toes

Multiplication and Division

Genius Level

1) Solve 1,230 x 2 = by breaking apart the 1,230.

2) Football teams usually have 11 players each. At a tournament I counted a total of 110 shoes. How many teams were there?

3) A chess set has 16 pawns, 2 kings, 2 queens, 4 bishops, 4 knights and 4 rooks. If Eric has 3 chess sets, how many total pieces does he have?

4) A room contains 48 spider legs, 42 grasshopper legs, 28 dog legs, and 12 duck legs.

The room has: ____ spiders, ____ grasshoppers, ____ dogs and ____ ducks.

5) 12 people have a total of 12 markers. If they **each** want to have 12 markers, how many more markers do they need to buy?

6) Skip count by 7s up to 77.

7) A farmer counted 56 total carrots. Assuming the rows are even, there are 4 possible arrangements of carrots. What could these arrangements be? (The first arrangement is completed for you.)

1 x 56 ___ **x** ___ ___ **x** ___ ___ **x** ___

8) An "exponent" looks like this: $\mathbf{5^4}$ and means **5 x 5 x 5 x 5**

How would your write this as an exponent: **3 x 3 x 3**

9) My friend and I *each* have a box of 36 crayons (they are arranged in rows of 6 crayons). If we combine our boxes, how many rows of 6 crayons can we make?

10) If you divide a dollar into 4 parts, then multiply one part by 4, what do you get?

16

Facts: Addition and Subtraction

Some teachers still give timed math fact tests because they were taught that students need to "just memorize" facts. But learning facts gets much more difficult when we don't teach the connections between them.

Just look at how many addition facts there are with the numbers 0-10:

	0	1	2	3	4	5	6	7	8	9	10
0	0	1	2	3	4	5	6	7	8	9	10
1	1	2	3	4	5	6	7	8	9	10	11
2	2	3	4	5	6	7	8	9	10	11	12
3	3	4	5	6	7	8	9	10	11	12	13
4	4	5	6	7	8	9	10	11	12	13	14
5	5	6	7	8	9	10	11	12	13	14	15
6	6	7	8	9	10	11	12	13	14	15	16
7	7	8	9	10	11	12	13	14	15	16	17
8	8	9	10	11	12	13	14	15	16	17	18
9	9	10	11	12	13	14	15	16	17	18	19
10	10	11	12	13	14	15	16	17	18	19	20

That's 121 different facts to memorize!

If you don't understand how these facts connect to each other, it is nearly impossible to memorize them all. That's like trying to memorize 121 different Latin words!

That is why teaching facts with only the "drill and kill" flashcard approach does not work well. It treats every fact as unique and rarely involves using different strategies to find an answer. Every addition fact *could* be solved by counting up by ones, but that way is slow and full of errors.

We can use flashcards smartly by grouping them into **Core Facts** and talking through strategies instead of setting time limits.

The end goal here is the same as always: Quickly find the correct answers.

The difference is that just memorizing facts without understanding them leads to a lot of mistakes. We want fast and accurate answers, but we also want kids to use good strategies when they get stuck.

And it is really boring to work on memorizing so many numbers. I like learning how numbers fit together, it makes math a lot more exciting!

We can master our math facts in two steps...

1. **Learn groups of facts that follow the same concept (for example: 10 + ____).**
2. **Use this knowledge to solve harder facts.**

Remember that strategies are slower at first but are a GREAT investment. With practice they become the fastest way to find the answers.

Timed Test Pitfalls

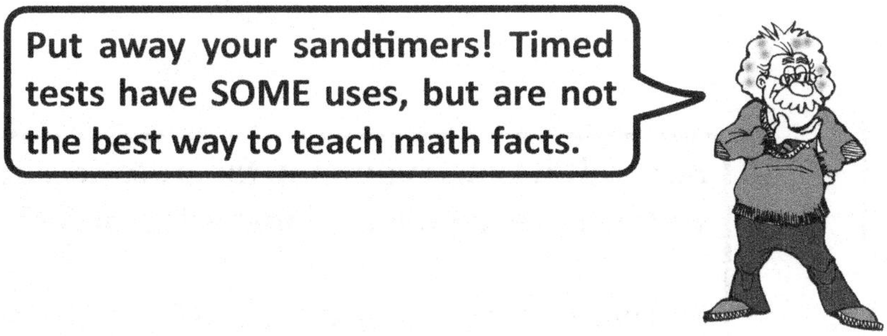

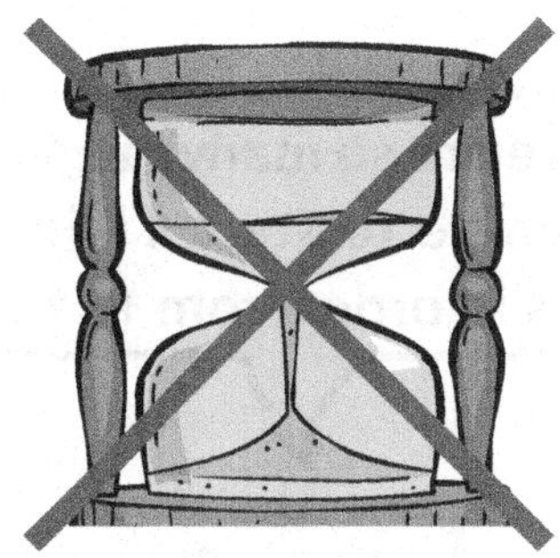

Less Information

Students usually show less work when they are timed. This means you won't have many clues to why a mistake happened.

Shallow Meaning

A higher score doesn't always mean that person is better at math. What if he skips around the paper to solve the easy questions first? Test-taking strategies are much different than number strategies.

Too Broad/Too Narrow

A test with only +3 addition facts (2+3, 5+3, 7+3, etc.) tells you little about what else a student knows. On the other hand, a random sample of *all* addition facts won't tell you what areas to target next.

Anxiety-Producing

Putting time pressure on people can make them anxious and more likely to fall back to an old crutch (like counting on their fingers for every problem).

Addition: Core Facts

But there are so many facts—I don't know where to start. I'll just teach the facts in order from +1 to +9.

Be careful! When you shop for groceries, do you buy everything in alphabetical order?

You would be running back and forth all day and your *butter* would melt by the time you made it to the *watermelons*!

This sounds like I'll be able to really prioritize certain groups so I can work smarter!

I'll call these the CORE facts, because they are the most important ones to learn. Look how much you'd know if you learned only the Core Facts!

Addition: Core Facts

	0	1	2	3	4	5	6	7	8	9	10
0	0	1	2	3	4	5	6	7	8	9	10
1	1	2	3	4	5	6	7	8	9	10	11
2	2	3	4	5	6	7	8	9	10	11	12
3	3	4	5	6	7	8	9	10	11	12	13
4	4	5	6	7	8	9	10	11	12	13	14
5	5	6	7	8	9	10	11	12	13	14	15
6	6	7	8	9	10	11	12	13	14	15	16
7	7	8	9	10	11	12	13	14	15	16	17
8	8	9	10	11	12	13	14	15	16	17	18
9	9	10	11	12	13	14	15	16	17	18	19
10	10	11	12	13	14	15	16	17	18	19	20

+0, +1, +2
Doubles
Combo 10
10 + ___

That's almost ⅔ of all the addition facts!

Here are some ways to teach the Core Facts of Addition:

+1, +2

These are the easiest addition facts to understand. Show how +1 or +2 relates to the counting sequence with the "how many now" routine.

 *How many?*

 *+1...How many now?*

 *+1...How many now?*

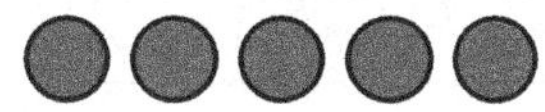 *+1...How many now?*

"See how adding 1 is just like counting?"

+0

Kids MUST understand what "zero" means before they can learn what it means to "add zero."

Don't just teach them a rule like "add zero means the number stays the same." Rules are easy to forget and we want people to remember *ideas* instead.

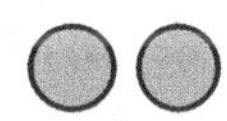 *How many?*

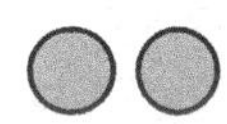 *+ 0...How many now?*

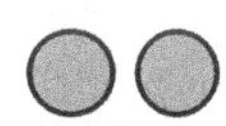 *+ 0...How many now?*

"Do you see what it means to add ZERO, or add nothing?"

Doubles

Doubles are easier to remember because they happen in real life objects (car wheels, fingers, eyes, etc). Practice talking about doubles when you see them.

Combo 10

These are all the addition facts that equal 10. They will take more work, but are important because of how often they are used.

"Ten-frames" help with these facts. It keeps the conversation about ten, and kids get practice hearing number pairs like 4+6, 9+1, 3+7...

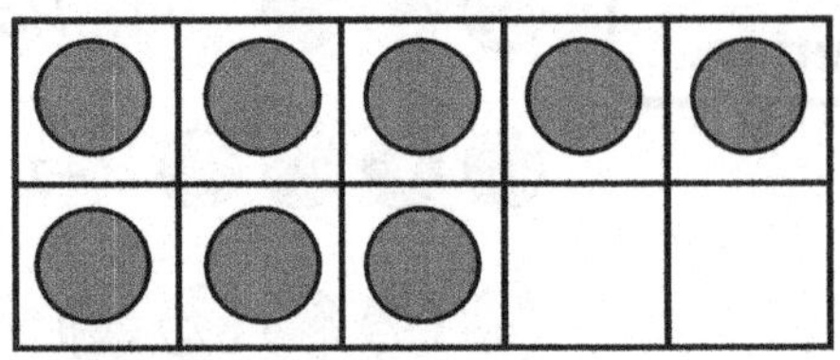

8 plus what equals 10?

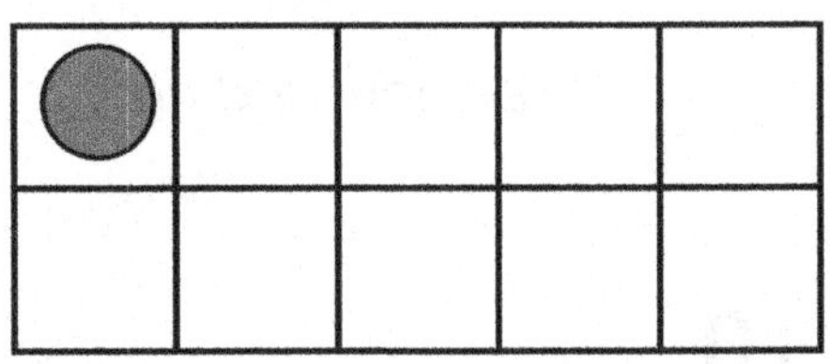

1 plus what equals 10?

Combinations of 10 are VERY important. They help when adding or subtracting large numbers...

13 + 8

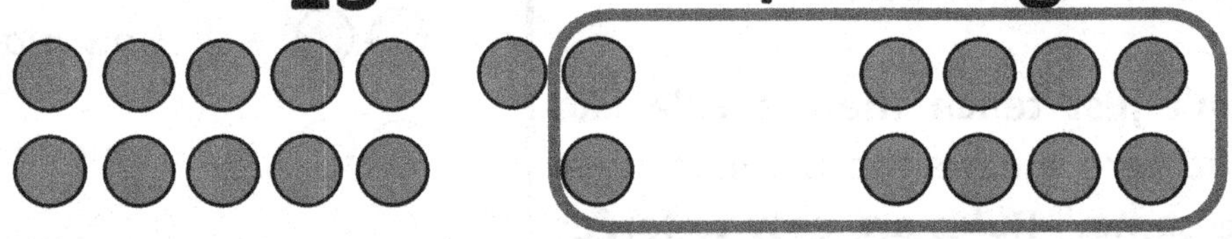

I know that 2 + 8 is 10, so it's like I have two 10s and a 1...

10 + ___ (teen numbers)

These facts help reinforce what "teen numbers" are. For these, it helps to bring out place value blocks to visualize things like "13 is really a 10 and 3."

At first, it is ok to start at the 10 and count up to find the answer. We want kids to realize the pattern themselves so they remember to apply it to harder problems in the future.

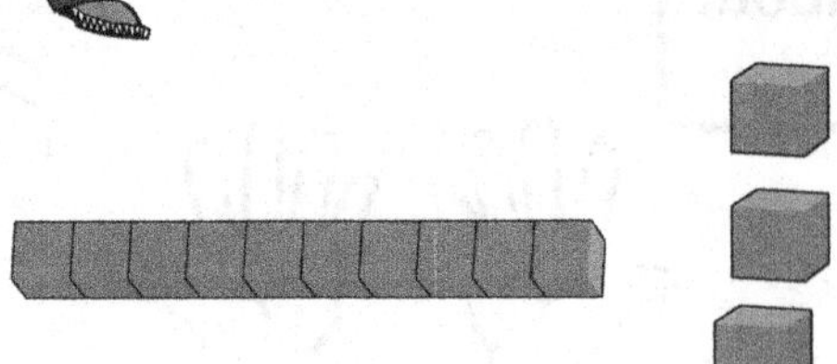

13 can be built from a 10 and ___?

Start at 10 and count the rest of the blocks...how many are there?

Addition: Derived Facts

	0	1	2	3	4	5	6	7	8	9	10
0	0	1	2	3	4	5	6	7	8	9	10
1	1	2	3	4	5	6	7	8	9	10	11
2	2	3	4	5	6	7	8	9	10	11	12
3	3	4	5	6	7	8	9	10	11	12	13
4	4	5	6	7	8	9	10	11	12	13	14
5	5	6	7	8	9	10	11	12	13	14	15
6	6	7	8	9	10	11	12	13	14	15	16
7	7	8	9	10	11	12	13	14	15	16	17
8	8	9	10	11	12	13	14	15	16	17	18
9	9	10	11	12	13	14	15	16	17	18	19
10	10	11	12	13	14	15	16	17	18	19	20

Near Doubles
Make 10

Near Doubles

1 or 2 away from a "doubles fact"

6 + 7 = ___

6 + 5   3 + 4   7 + 5

Make 10

Break apart a number to make a group of 10

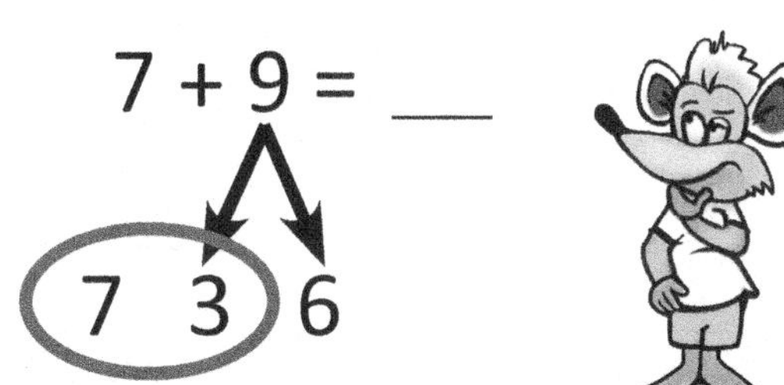

The goal is to answer each fact mentally and automatically within 3-5 seconds. Strategies like "Make 10" are meant to be done in your head and will take a little longer at first, but with practice they will quickly become second nature!

Making Practice Fun

Flashcards are not very fun. With a little preparation, you can make games to practice the facts you are working on!

Memory

Match facts with the answer (or with another fact that has the same answer).

War

Flip two cards and say the sum. Get the answer right to keep the cards until the deck is gone.

Bingo

You can use a numbered board for different math facts. Choose a fact group and call out numbers. The child puts a marker on the answer. *See page 339 for a blank Bingo board you can use!*

For Example (**Doubles**):

Say "2" (*Marks* 4)

Say "5" (*Marks* 10)

Say "9" (*Marks* 18)

3	1	12	5	0
14	18	7	4	8
17	15	X	19	13
9	2	7	6	18
0	11	16	10	2

For more math fact games, check out *Math Fact Fluency: 60+ Games and Assessment Tools to Support Learning and Retention* by Jennifer Bay-Williams and Gina Kling.

Subtraction

Once we know our addition facts, we can think about subtraction. Thankfully, these facts have a special relationship with addition.

Numbers that go together like (15, 7, 8) are called "Fact Families." They help bridge the gap between addition and subtraction.

Addition Fact

8 + 7 = 15

Subtraction Fact

15 - 8 = 7

At school we use something called "Number Bonds" to learn about fact families!

Here are three other subtraction strategies you can practice:

Count Back

Best for subtracting small amounts.

17 - 4 12 - 2 19 - 3

Count Up

Best when numbers are close to each other

17 - 15 12 - 9 19 - 17

Compare

Find the "difference" between the numbers, as in the gap between them. This works when there is some distance between the numbers. A number line can help visualize this distance and keep track of how far you've gone.

17 - 9

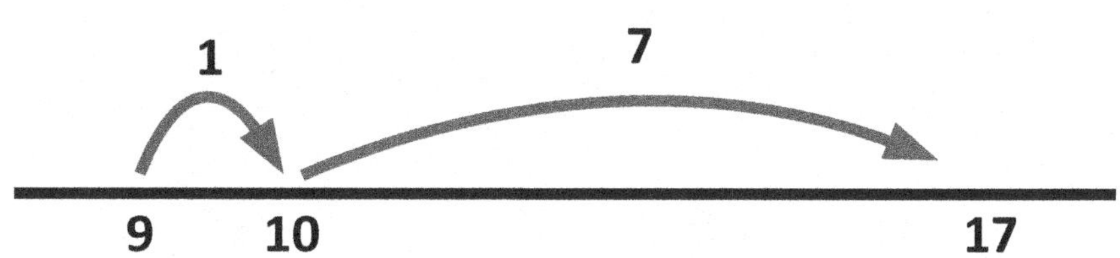

I move up to 10 *then jump to 17*

that means a total of 8 between them

Math Facts: Addition/Subtraction

Level 1

1) How many oranges?

(Plus 1) how many now?

\+ 

(Plus another 1) how many now?

\+ 

2) What is 3 + 0?   15 + 0?   7 + 0?
What pattern do you notice about adding zero?

3) When you count, what number comes after 7? What is 7 + 1?
When you count, what number comes after 9? What is 9 + 1?

What pattern do you notice about adding 1?

4) Circle the numbers that are doubles?

4   5   2   8   12

5) Draw a picture to show why 3 + 4 is the same as 4 + 3.

Math Facts: Addition/Subtraction

Level 2

1) Circle all of the subtraction problems that you should "count back" to find the answer.

208 - 3   178 - 170   19 - 17   13 - 1   1,980 - 0

2) Circle all of the addition facts that are "Near Doubles."

9 + 8	2 + 9	2 + 1	3 + 4	5 + 7
7 + 2	3 + 5	0 + 10	8 + 7	5 + 5

3) "Number Bonds" are a way to show fact families, or the relationship between adding and subtracting numbers. Complete the rest of these Number Bonds:

4) What kind of multiplication math facts will be the same as Addition's "Doubles"?
Hint: Doubles facts look like this: 7+7, 3+3, 2+2, etc.

x1   x2   x3   x4   x5

5) Use the "Make 10" strategy to solve these addition facts. Describe *how* you are making 10 to help solve them.

5 + 8   3 + 9   7 + 5   19 + 5   27 + 4

Math Facts: Addition/Subtraction

Level 3

1) Only one of these Number Bonds is used correctly. Find the two **incorrect** Number Bonds and explain why they are wrong.

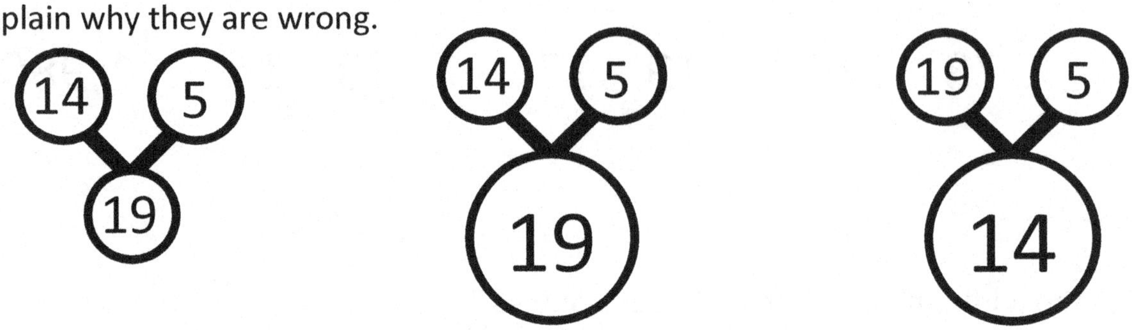

2) What are these sums?

9 + 8	2 + 9	2 + 1	3 + 4	5 + 7
7 + 2	3 + 5	0 + 10	8 + 7	5 + 5

3) Predict which group of Multiplication facts will be important to teach first (*x0, x1, x2, x3, etc*). Which group of Multiplication facts do you think are the hardest to understand?

4) Look at these Number Bonds that have variables instead of numbers.

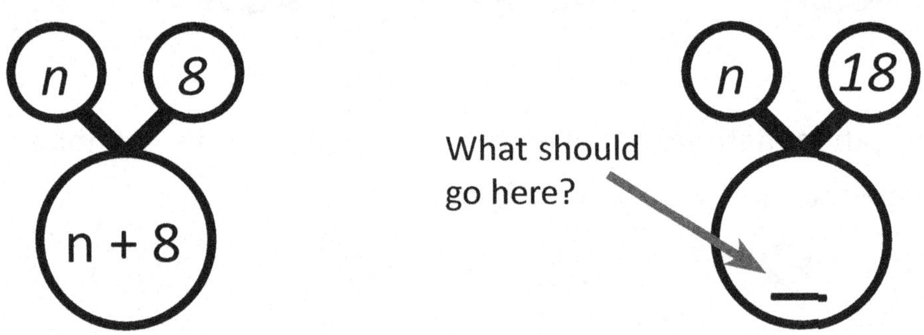

5) Explain why the "Combinations of 10" and "Make 10" facts are so important.

Math Facts: Addition/Subtraction

Genius Level

1) Complete these very difficult number bonds:

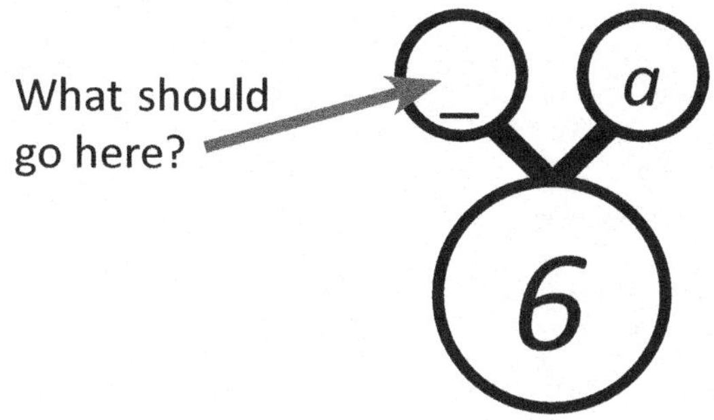

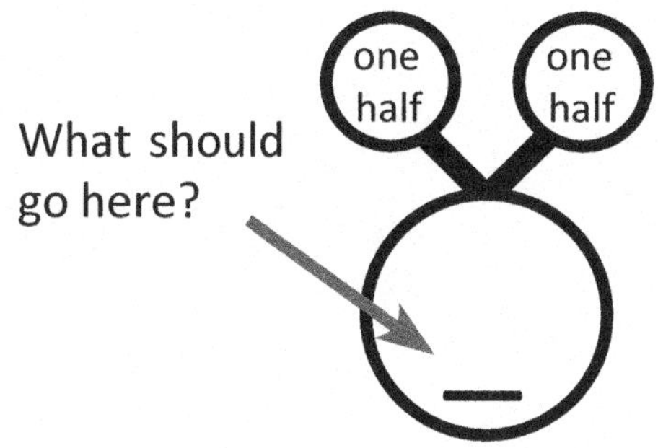

2) Complete this number bond:

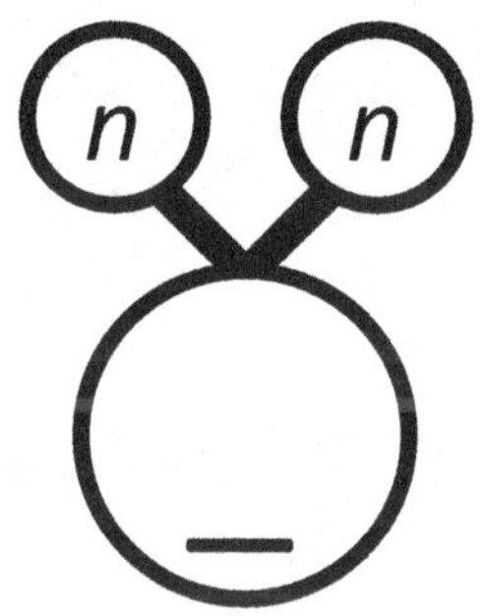

3) Skip count by 12s until you reach 144.

4) Skip count by 9s **backwards** from 90 to 0.

5) Complete this number bond:

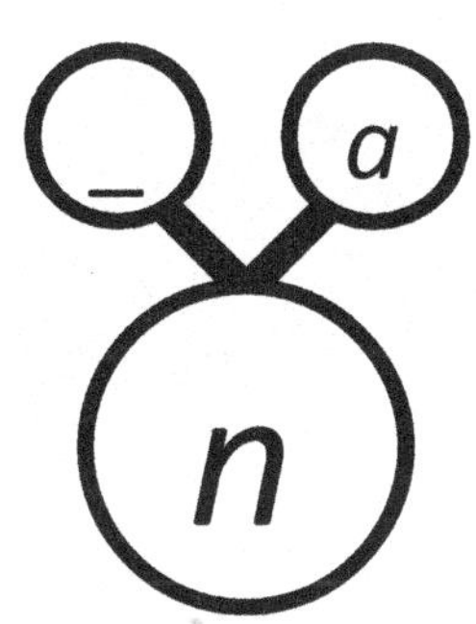

17

Facts: Multiplication and Division

Thankfully, a lot of the ideas from the Addition Facts chapter apply here as well!

Right! Like...

- The goal is to know all of our facts automatically.
- Some groups of facts are easier to learn than others.
- Strategies and games make learning facts more fun than just memorizing numbers.

My teacher just gave me a list of "tricks" to remember multiplication facts. Isn't that good enough?

Be careful with tricks! They usually stop you from thinking about what the numbers actually mean. However, it's ok to use them sometimes as long as you understand *why* they work.

Rules aren't bad, but they are NOT a substitute for learning important math ideas like "how numbers relate to each other."

So maybe the best "trick" of all is using number sense to work with numbers flexibly and efficiently!

My sister is such a show-off.

Multiplication: Core Facts

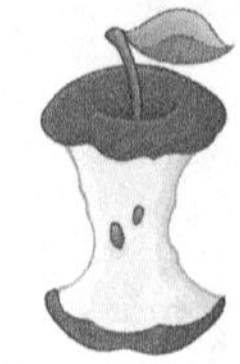

x2 (doubles)
x5
x10
x1
x0
Squares

	0	1	2	3	4	5	6	7	8	9	10
0	0	0	0	0	0	0	0	0	0	0	0
1	0	1	2	3	4	5	6	7	8	9	10
2	0	2	4	6	8	10	12	14	16	18	20
3	0	3	6	9	12	15	18	21	24	27	30
4	0	4	8	12	16	20	24	28	32	36	40
5	0	5	10	15	20	25	30	35	40	45	50
6	0	6	12	18	24	30	36	42	48	54	60
7	0	7	14	21	28	35	42	49	56	63	70
8	0	8	16	24	32	40	48	56	64	72	80
9	0	9	18	27	36	45	54	63	72	81	90
10	0	10	20	30	40	50	60	70	80	90	100

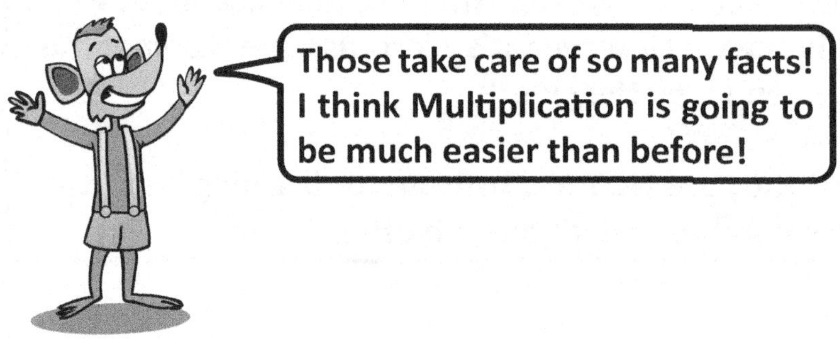

Here are some ways to teach the Core Facts of Multiplication:

2s (Doubles)

Show what it means to "Double" a number. Story problems that involve natural rows of 2 can be a good way to visualize the doubling. Show how "times 2" means "two rows" or "two groups" of a number.

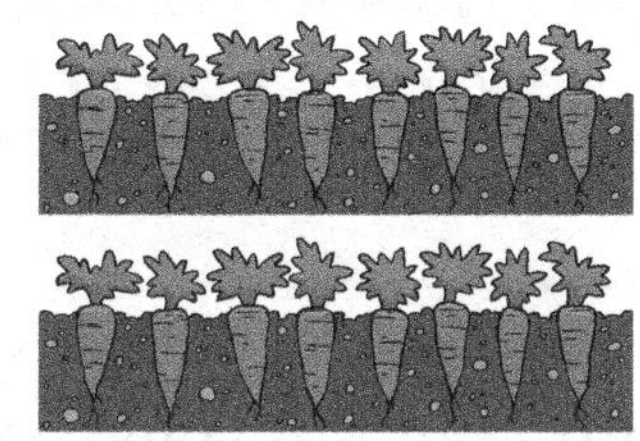

> Skip counting groups (2, 4, 6, 8...) is ok at first, but it is a more basic strategy that will eventually slow children down. It is one tool, but should not be the only tool!

Help children see *why* **x2** and **Doubles** facts are the same. Once they see that connection, they will automatically know their **x2** facts!

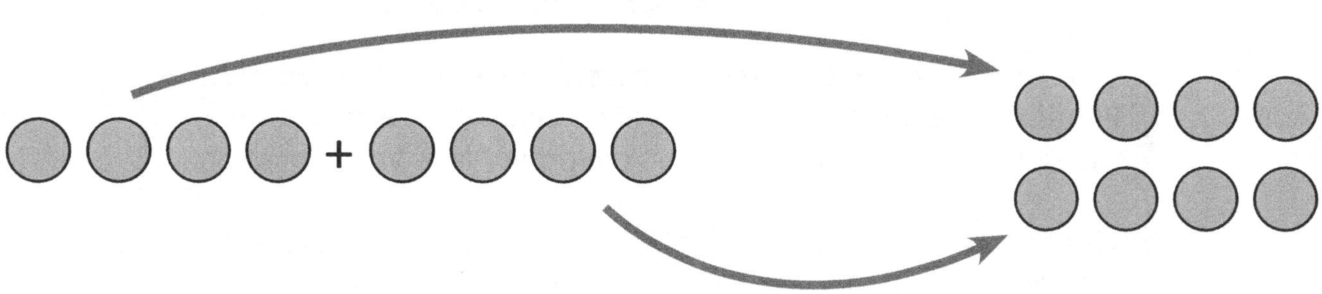

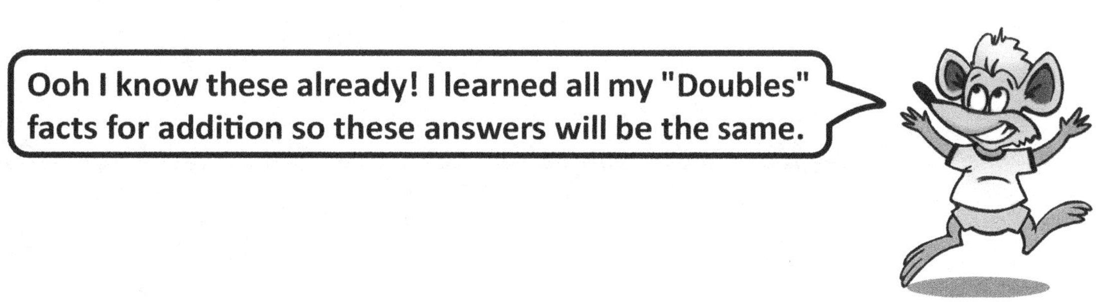

10s

"Tens" are another easy fact to start with. The key here is **Place Value,** so children should have experience with that before working on the **10s** facts.

I know that "times" means "groups of." Four groups of ten is just like four tens. Four tens...forty!

I can double check my answer with Place Value blocks or by skip-counting.

4 x 10

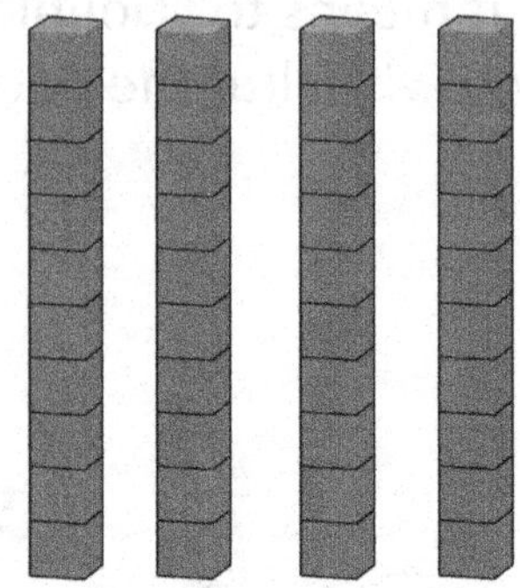

"10...20...30...40"

It is tempting to just say, "Times 10 means add zero," but this could backfire later.

Look what happens when you "add zero" to these x10 questions. It doesn't work anymore!

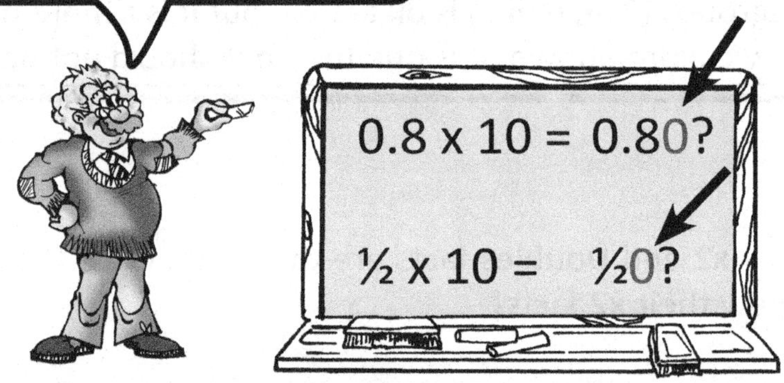

By relating "times 10" to Place Value, children see that it means the same as shifting the numbers over one Place Value location.

This also translates to Division *much* better than the "add zero" rule, because to divide you just do the same thing backwards.

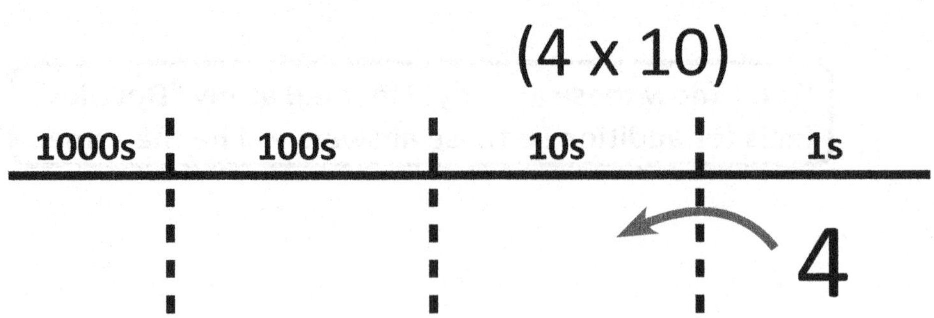

1s

These are easy to master once children understand "times" means "groups of." It is easy to understand what *one group* of a number is.

3 x 1

Three groups of **One** apple

1 x 3

One group of **Three** apples

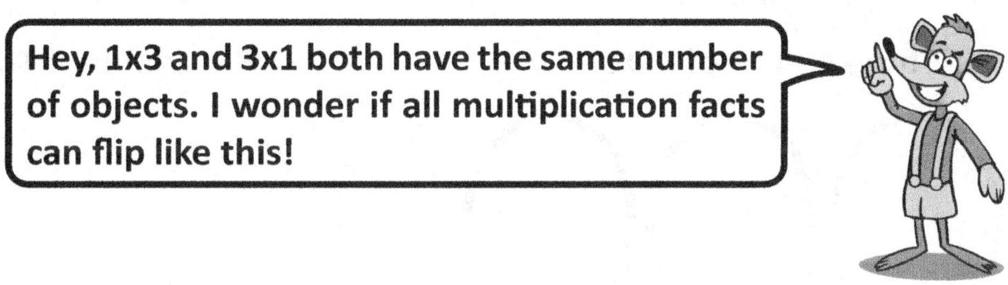

5s

Five is an important benchmark in our counting system, which makes learning **x5** facts a priority. Here are a few ways to teach and remember them:

Skip count

(5...10...15...20...etc.)
This is a little more acceptable for 5s because it can happen fairly quickly, but it is still a more basic way of thinking about numbers.

Think 10

(5 groups is half of 10 groups, so I can think x10 and split it in half)

This keeps children thinking about relationships between numbers. It also applies to much larger numbers *(e.g. 5 x 820).*

5 x 6

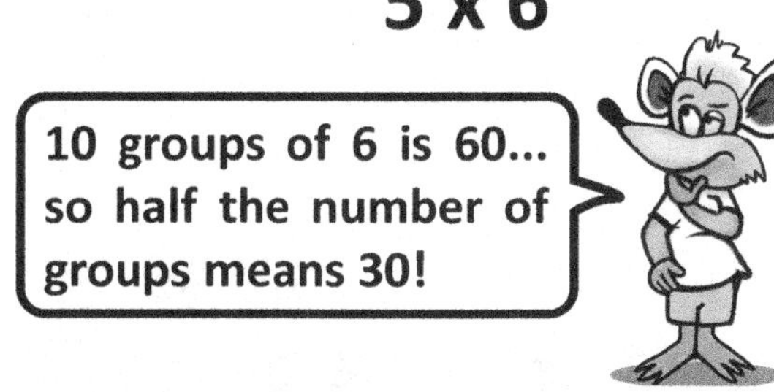

Use Benchmarks

(5 x 5 is 25, so 7 x 5 means two more groups...25 + 5 + 5 = 35)
Remembering benchmarks like 5 x 5 (Square Fact) gives us a starting point for adding or taking away groups.

0s

I bet you are surprised we didn't START with x0 facts. But remember, I'm still an abstract idea because it is harder to visualize "having zero" of something.

I know the "everything times zero = zero" rule is tempting, but in the long run it will help to understand what x0 really means. Zeros just want you to get to know them better...

3 x 0	**0 x 3**
Three groups of **NO** apples	**NO** groups of **Three** apples

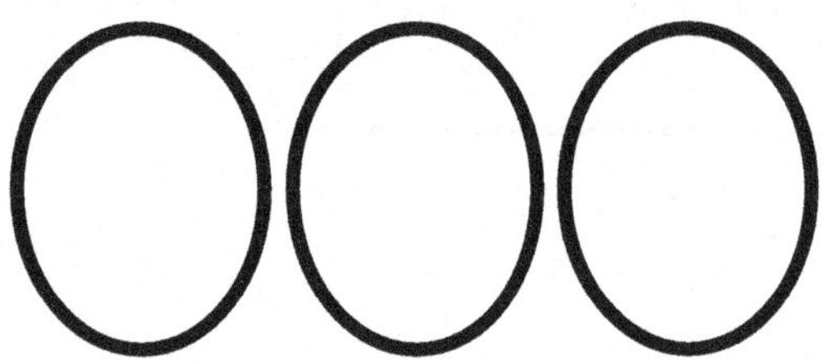

It only takes a little extra time to teach exactly what __ **x 0** and **0 x** __ really mean. Rules are easy to mix up and forget. Number Sense will never let you down because its patterns are predictable and built on each other.

Squares

Squares are important benchmarks that help with harder multiplication facts (*for example, knowing 7 x 7 = 49 will help us quickly find 7 x 8)*. These facts also help when kids start learning about exponents or finding the area of square shapes.

Dot patterns like this help show what it means to "square" a number.

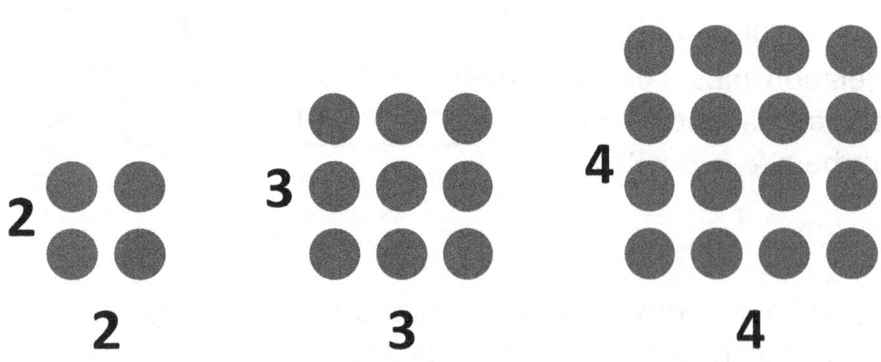

Multiplication: Derived Facts

Look how few facts we have left! Knowing our Core Facts really took care of a lot of Multiplication!

Let's look at the "Derived Facts." These use strategies to build on the foundation we have already learned.

Double-Double (4s)
Add/Subtract a Group
Break Apart

	0	1	2	3	4	5	6	7	8	9	10
0	0	0	0	0	0	0	0	0	0	0	0
1	0	1	2	3	4	5	6	7	8	9	10
2	0	2	4	6	8	10	12	14	16	18	20
3	0	3	6	9	12	15	18	21	24	27	30
4	0	4	8	12	16	20	24	28	32	36	40
5	0	5	10	15	20	25	30	35	40	45	50
6	0	6	12	18	24	30	36	42	48	54	60
7	0	7	14	21	28	35	42	49	56	63	70
8	0	8	16	24	32	40	48	56	64	72	80
9	0	9	18	27	36	45	54	63	72	81	90
10	0	10	20	30	40	50	60	70	80	90	100

Double-Double (4s)

By now, kids should be able to "Double" a number automatically. Connect "Times 4" to this by showing what it means to double a number and double it again.

3 x 4

I know my Addition Doubles facts really well, so if "times 4" means to Double-Double, I can do those too!

3 doubled is 6, and 6 doubled is 12!

Add/Subtract a Group

This means using a nearby fact to quickly solve a harder one.

Not sure what 8x6 is? Well, you know that 8x5 = 40, so one more group is 48.

Be sure to show this idea VISUALLY at first so kids can see why adding/subtracting a group works.

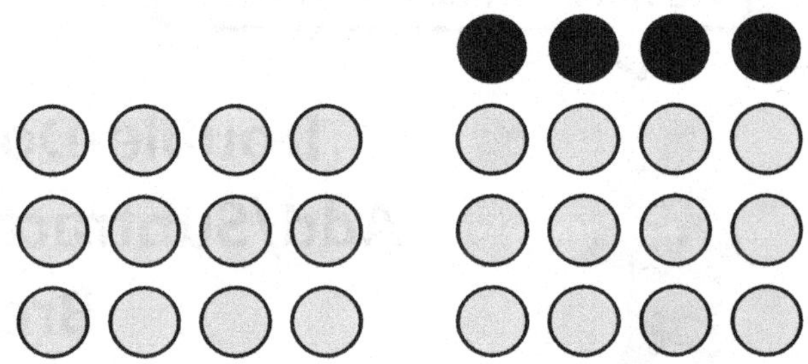

3 x 4

4 x 4
(one more group)

I used this idea to learn my x9 facts!

9 x 7

I'm going to pretend there are 10 groups because that's an easier number to work with.

10 x 7 is 70, then I just need to subtract one group of 7. That leaves me with an answer of 63! After practicing this a few times, I can do it all in my head quickly.

Pay special attention to adding/subtracting a group for "Square Facts." This connects to what kids already know, and these facts are common for math problems.

Break Apart

This is the most complicated and advanced strategy, but it can be used for almost any multiplication fact. It involves breaking apart numbers to make them easier to work with.

3 x 4

3 groups of 2

3 groups of 2

3 groups of 2 = 6
3 groups of 2 = 6
(Total = 12)

It helped me to see visuals for this while I was still learning how to break apart numbers.

3 groups of 4

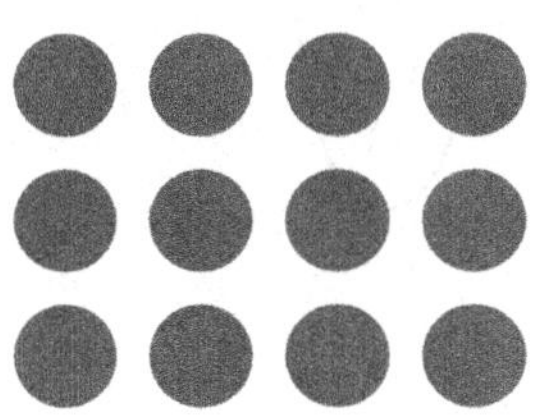

OR

3 groups of 2
+
3 groups of 2

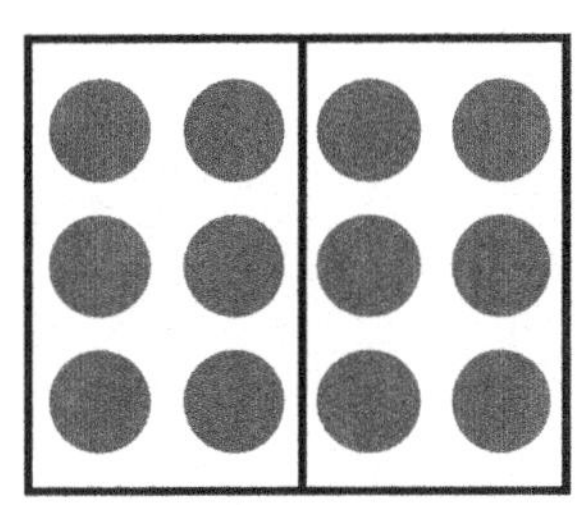

This is an advanced idea. Kids will need a lot of practice and examples to master it, but "Break Apart" will go a long way towards helping them become flexible with numbers.

"Break Apart" is a universal strategy. It will help with lots of math problems like fractions, area, algebra, and more!

Division

Just like Addition facts help with Subtraction, Multiplication facts help with Division.

Numbers that go together like this (3, 4, 12) are called "Fact Families." They help bridge the gap between Multiplication and Division.

Multiplication Fact

3 x 4 = 12

Division Fact

12 ÷ 4 = 3

Arrays can be another strategy to work with division facts. For example, showing that 12 can be broken into 4 equal groups of 3.

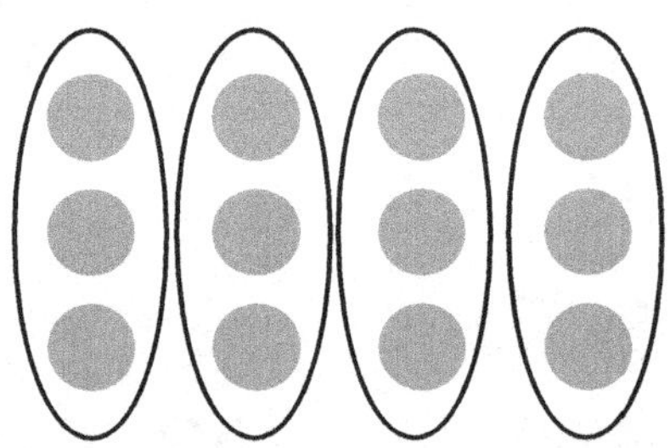

Any of the activity ideas on the next page will apply to Division as well. Once your child understands Multiplication, try some more challenging Division problems!

Making Practice Fun

The "Addition Fact" game ideas work for multiplication, too! Practicing facts this way will be more fun and useful than flashcards.

Memory

Match facts with the answer (or with another fact that has the same answer).

War

Flip two cards and say the product (the numbers multiplied). Get the answer right to keep the cards until the deck is gone.

"16!"

Bingo

You can use a numbered board for different multiplication facts. Choose a fact group and call out numbers. The child puts a marker on the answer. *See page 339 for a blank Bingo board you can use!*

For Example (**2s**):

Say "2" (*Marks* 4)

Say "5" (*Marks* 10)

Say "7" (*Marks* 14)

"2s Board"

6	2	12	8	14
(14)	18	9	(4)	8
17	16	X	18	(10)
8	2	6	12	18
2	8	16	10	2

Note: The bingo board will look different for some groups of facts. You can quickly write the numbers after deciding on your fact group (e.g. Pick "Squares" and write 36...25...81, etc).

For more math fact games, check out *Math Fact Fluency: 60+ Games and Assessment Tools to Support Learning and Retention* by Jennifer Bay-Williams and Gina Kling.

Math Facts: Multiplication and Division

Level 1

1) What are the Doubles of these numbers?

6 **3** **1** **9** **7**

2) What are these numbers times 10?

1 **5** **17** **92** **4**

3) Explain how you can use the fact 6 x 10 = 60 to tell you what 6 x 5 is.

4) What does it mean to multiply a number by 1? (Think groups.)

5) Doubling a number is the same as multiplying by ____.

Math Facts: Multiplication and Division

Level 2

1) What do you get when you double these numbers twice? (Double-Double)

8 **5** **0** **6** **3**

2) Quickly say the answers without skip-counting.

5 x 2	5 x 7	6 x 5	1 x 5	5 x 0
5 x 8	5 x 5	5 x 3	5 x 4	5 x 9

3) Complete these facts:

6 x __ = 30	__ x 7 = 56	48 ÷ __ = 6
64 ÷ __ = 8	4 x 9 = __	__ x 7 = 21

4) Finish this sentence: **x4** facts are ______ as big as **x2** facts.

5) Doubling a number twice is the same as multiplying by ____.

Math Facts: Multiplication and Division

Level 3

1) Doubling a number two times is the same as multiplying by __4__
   Doubling a number three times is the same as multiplying by ____.
   Doubling a number four times is the same as multiplying by ____.
   Doubling a number five times is the same as multiplying by __32__

2) Quickly say the answers without skip-counting.

2 x 2	7 x 7	6 x 6	1 x 1	4 x 4
9 x 9	3 x 3	0 x 0	5 x 5	8 x 8

3) Explain how you would solve these using the "Add/Subtract" a group strategy. Be sure to say what fact you used to find the final answer.

For Example: **3 x 4** *(3 x 3 is 9, add one more group of 3 for **12**)*

7 x 8	9 x 8	6 x 7	4 x 3	8 x 9
6 x 5	8 x 6	7 x 9	5 x 3	6 x 7

4) Draw a picture to show why 3 x 2 is the same as 2 x 3:

5) Use the "Break Apart" strategy to solve this problem: **6 x 8**

Math Facts: Multiplication and Division

Genius Level

1) 32,482,612 x _____ = 32,482,612

   32,482,612 x 0 = ______________

2) 5 x 7 = 35 so 35 ÷ 7 = 5

   a x b = c so ___ ÷ ___ = a

3) 7 x n = 140 so 140 ÷ ___ = ___

4) 1 x 1 x 1 x 1 x 2 x 1 x 1 x 1 x 1 = _______

5) 412 x 500 x 100 x 42 x 0 = ________

18

Fractions

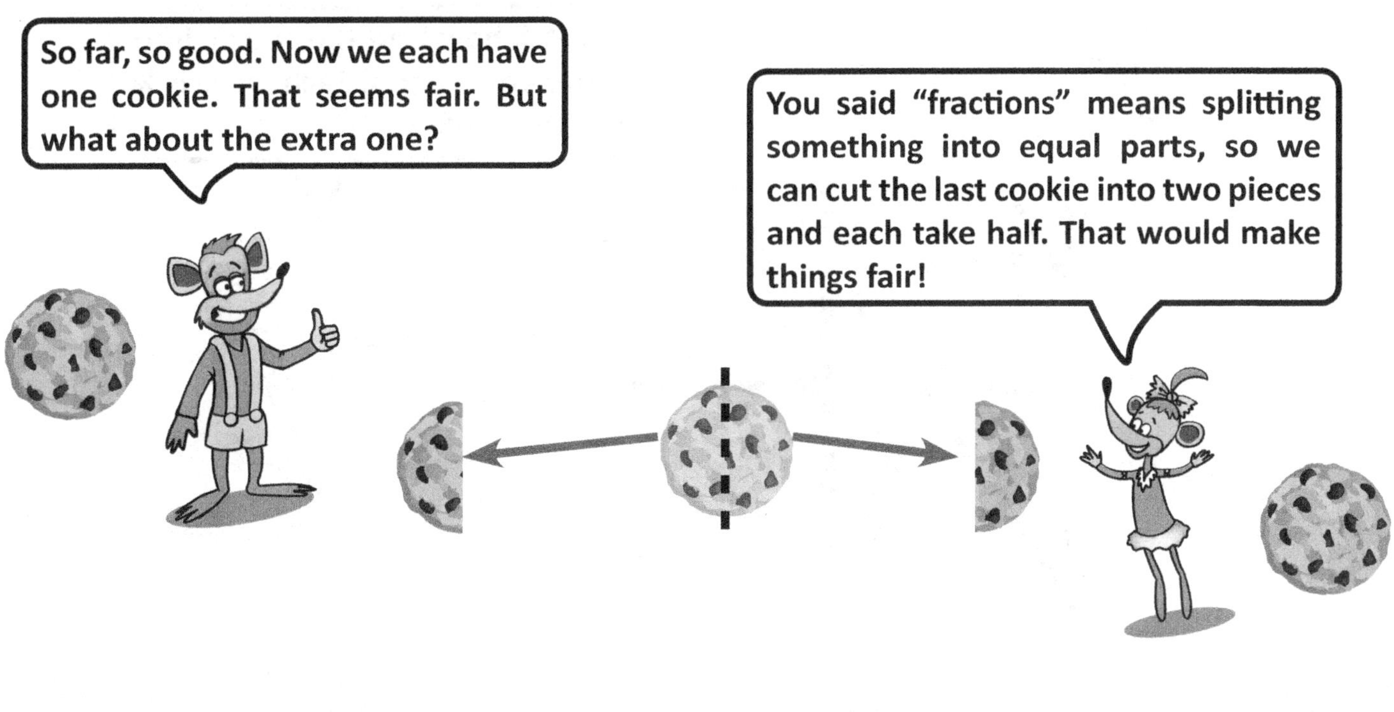

Learning fractions will help you do a lot of things such as splitting items fairly, cooking, and even understanding decimals and percents! Sometimes it is easier to think about splitting ONE thing. Earlier we split a cookie into two pieces, each making one half, but what if we had to split a *chocolate bar* in half instead?
I would pretty much do the same thing, but instead of cutting a circle in half, I would just cut the rectangle in half.
That seems easy, but what if we had to share a cookie with 4 people instead of 2?
Watch how I cut the cookie into four equal parts. Each piece is ¼ of the cookie.
¼
¼
¼
¼

The In-Between Numbers

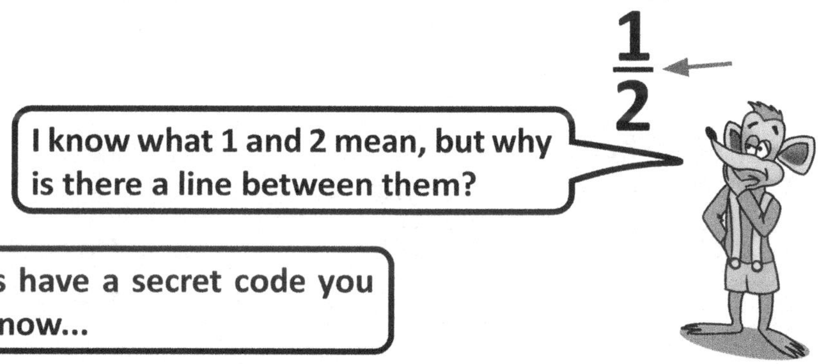

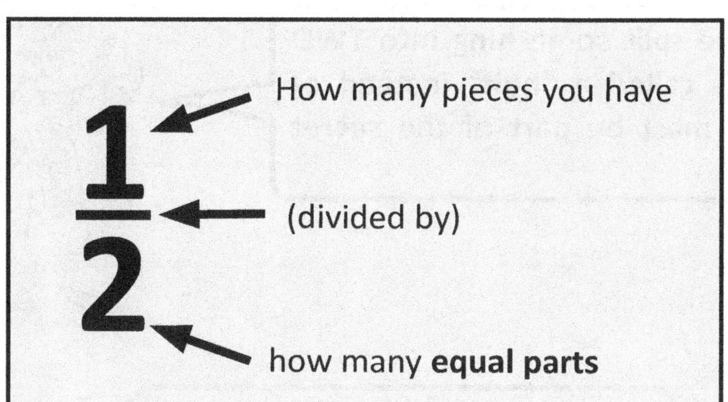

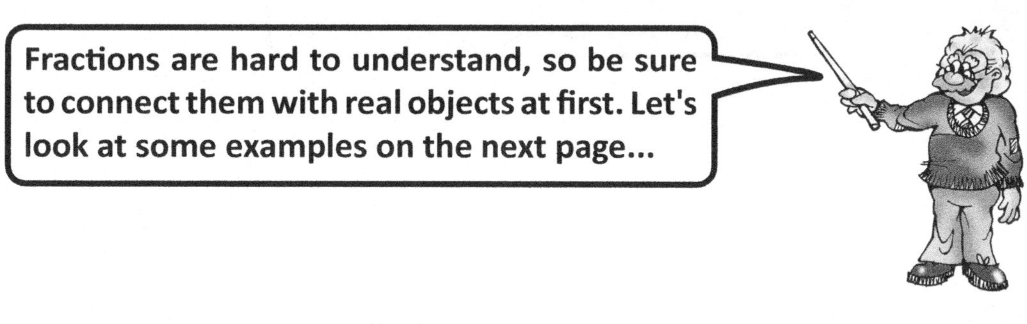

One Fourth:
A cookie is split into **4** equal pieces, you get **1** piece.

$$\frac{1}{4}$$

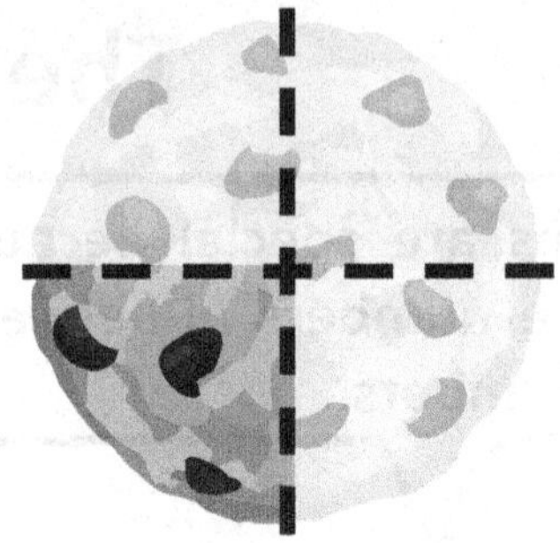

Two Fourths:
A cookie is split into **4** equal pieces, you get **2** pieces.

$$\frac{2}{4}$$

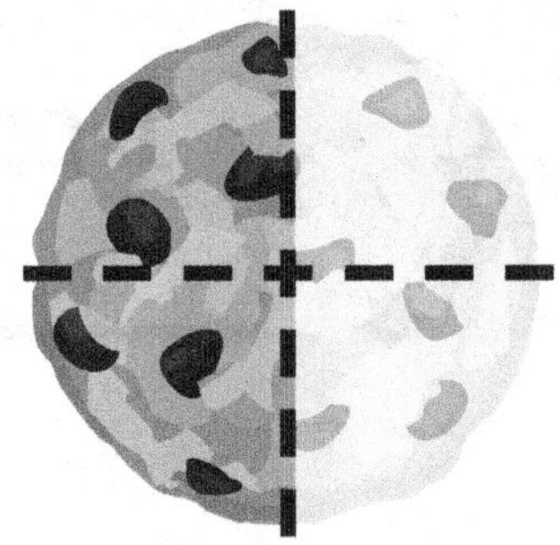

One half:
A cookie is split into **2** equal pieces, you get **1** piece.

$$\frac{1}{2}$$

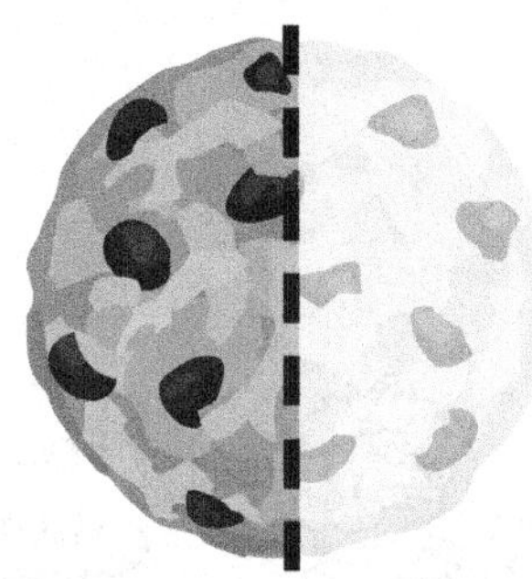

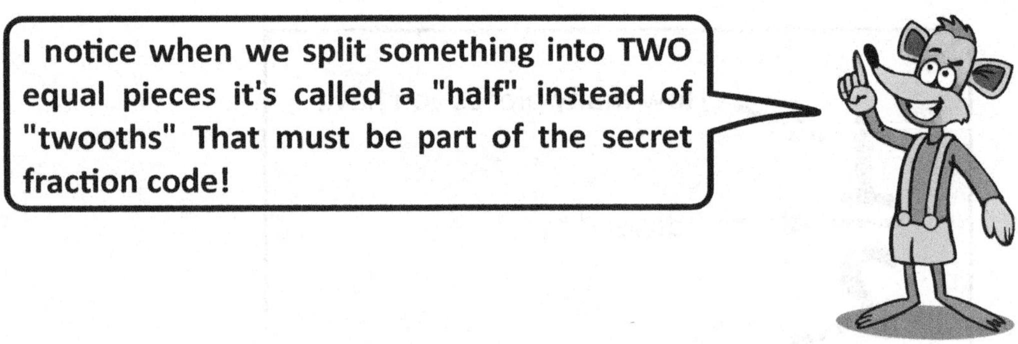

Different Name, Same Amount

I noticed those last two examples looked the same but had different fraction names.

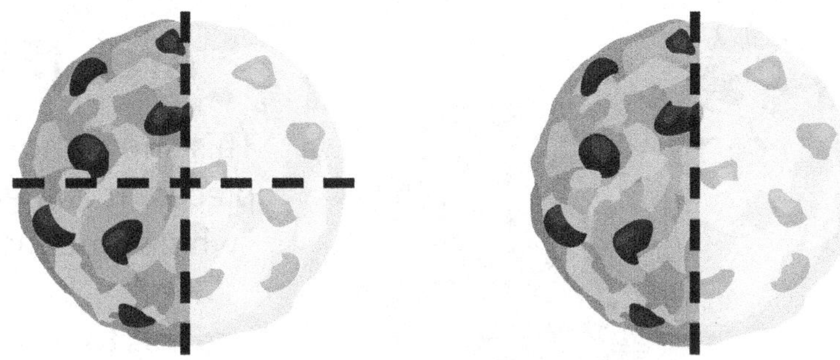

$\frac{2}{4}$ $\frac{1}{2}$

Those are called "equivalent fractions" and it means the amounts are the same even though we split the cookie up in different ways.

But be careful...we can only compare fractions when they come from things that are the same size!

That makes sense, because ½ of a cookie is a lot smaller than ½ of a pizza!

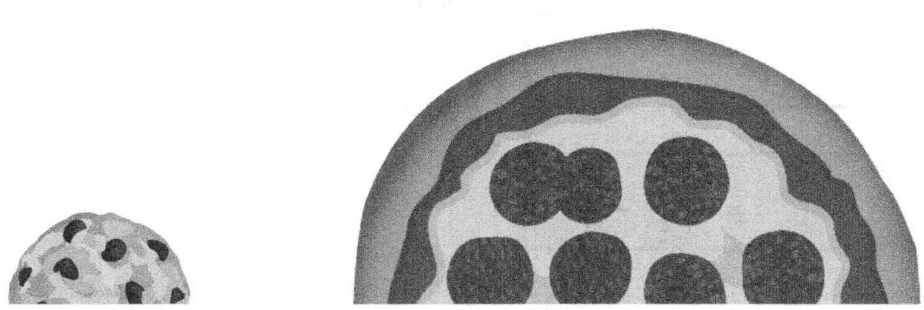

Words matter! NEVER ask someone to *reduce* a fraction, because "reduce" means to make smaller. Instead, we ask people to **simplify** fractions.

½ is not smaller than 2⁄4, but it is simpler!

Three Ways to Fraction

There are three different ways to think about fractions...

Area (Pizza)

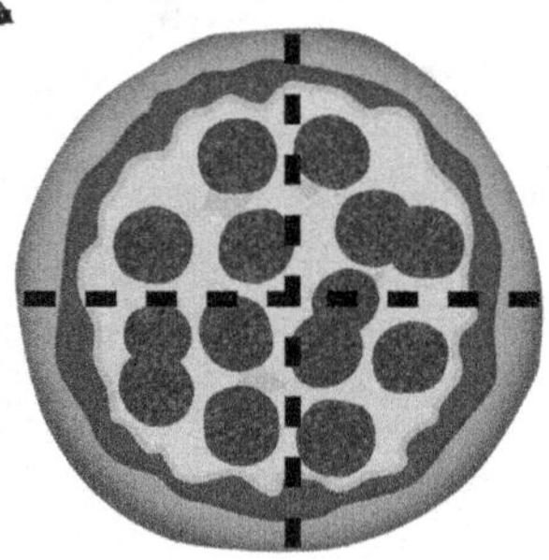

This shows how one object can be equally split into different pieces. It is a good starting point because it is clear that the "whole" means the entire pizza.

Use this to show what fractions like "half" and "fourths" mean.

This type of fraction is usually not the best for comparing different fraction sizes. You might see that $\frac{3}{4}$ pizza is bigger than $\frac{2}{3}$ pizza, but it doesn't show the *relationship* between the sizes.

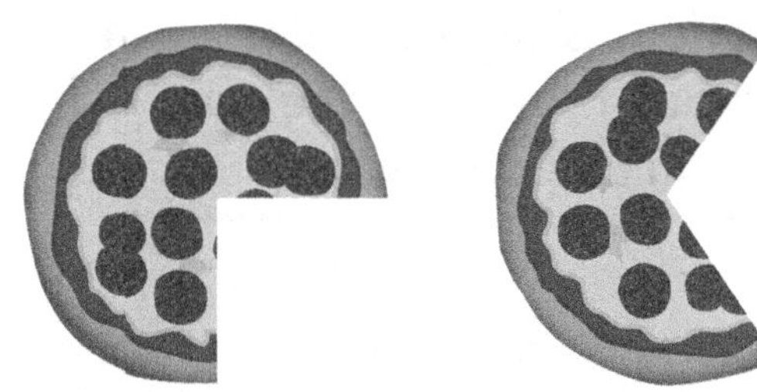

Lines

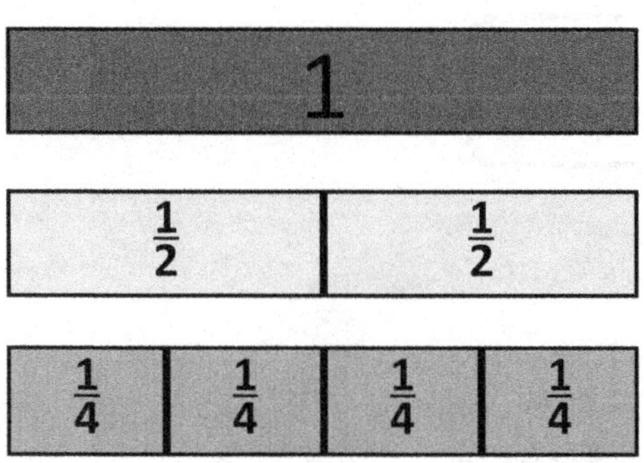

These help show the relationship between different fraction sizes. For example, notice how clearly this shows that $\frac{1}{2} + \frac{1}{2} = 1$ or that $\frac{1}{4} + \frac{1}{4} = \frac{1}{2}$.

These are easy to make at home by folding paper strips and labeling the pieces.

A rectangle by itself doesn't show the "whole" like a circle does. You will always need to keep the "1" in mind, otherwise the child will wonder "$\frac{1}{2}$...of what?"

See page 340 for fraction strips you can copy and use!

Objects

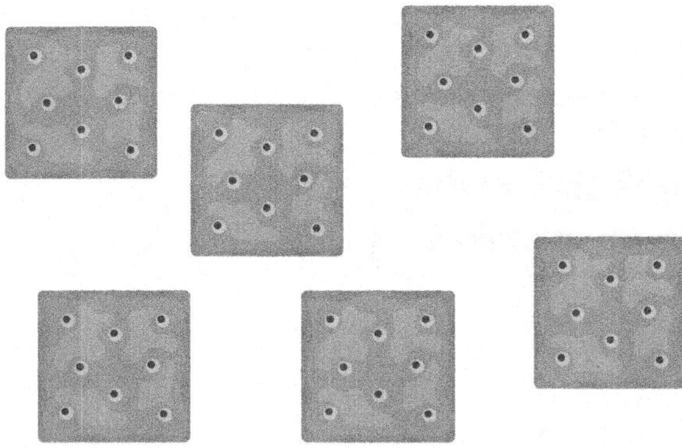

This can connect the ideas of "fractions" and "division." It helps show the idea of fairness and can be acted out with real-life items.

This is the hardest type of fraction for young children to understand. It makes sense to think about a pizza or rectangle as the "whole," but it's harder to see the "whole" as a group of objects. This is especially true when the number of objects in a group changes.

Combining Fractions

Fractions can be a little tricky to add, subtract, multiply, and divide. It helps to think about what fractions *mean* to avoid any common mistakes.

The Big Error:

$$\frac{1}{4} + \frac{1}{2} \rightarrow \frac{2}{6}$$

That looks like the person wasn't thinking about what the fraction numbers mean. The bottom part of a fraction means "how many equal parts." It isn't a regular number, so we can't add it like the other numbers.

Adding the bottom parts would be like adding these two pieces of fruit. The answer would not make any sense!

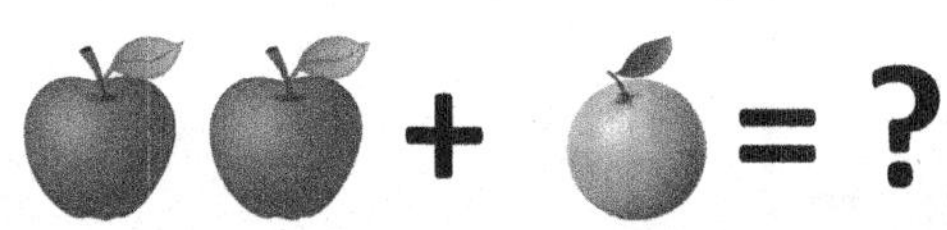

$$\frac{1}{4} + \frac{1}{2}$$

Here is where fraction strips can be useful. Always have tools like this ready when kids are first learning about fractions.

Here they show what ½ + ¼ looks like compared to the whole...

...and also show how ½ + ¼ is the same as ¾.

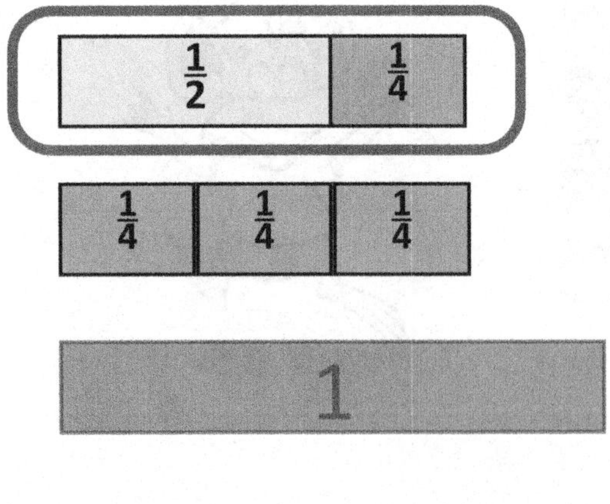

Grouping/Dividing Fractions

Multiplying and dividing fractions? Oh no, this sounds really difficult!

It can be challenging, but it will be much easier if you first think about what fractions mean!

$\frac{1}{4} \times 3 =$  Three groups of ¼ piece

$\frac{1}{4} \times \frac{1}{2} =$ 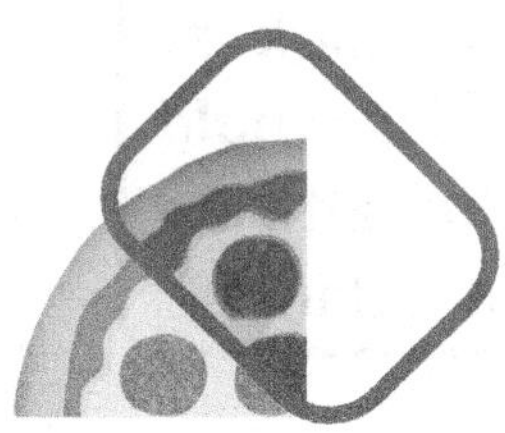 a HALF group of ¼ piece

$\frac{1}{4} \div 2 =$  ¼ piece, split into two equal parts

I know I will learn a lot more in school about this, but it helps to understand the basics!

Balance the Scale!

Here is another fun thing we can do with fractions. How many cookies would I need to add to the right side of the scale to balance it?

That is really easy. All I would have to do is add $\frac{1}{2}$ of a cookie to the right side, then there would be the same number of cookies on both sides!

Excellent thinking! Here is a more difficult fraction problem. Look at this scale. How many cookies would you have to move from the left side to the right side to make it balance?

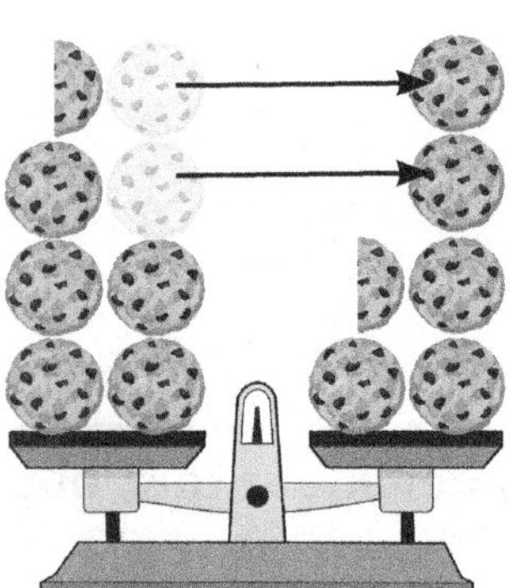

That *is* harder. If I move one cookie from the left to right side, then there will be $6\frac{1}{2}$ and $4\frac{1}{2}$. But those aren't the same...

...so I will take one more cookie from the left side to get $5\frac{1}{2}$ and $5\frac{1}{2}$!

Fraction Benchmarks

Here are some important fraction benchmarks to remember:

Visuals	*Fraction*	*Words*
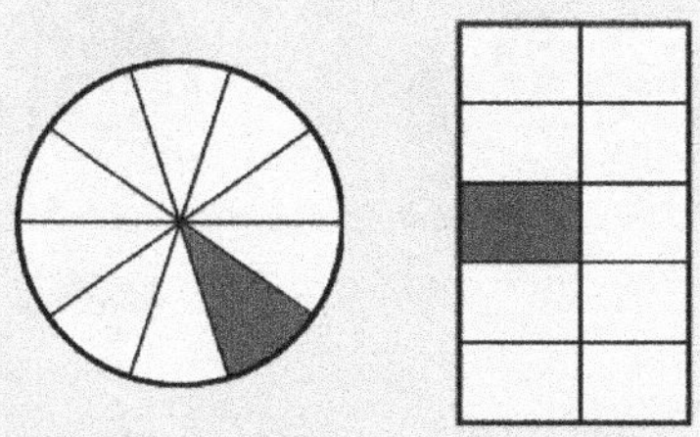	**1/10**	**One tenth**
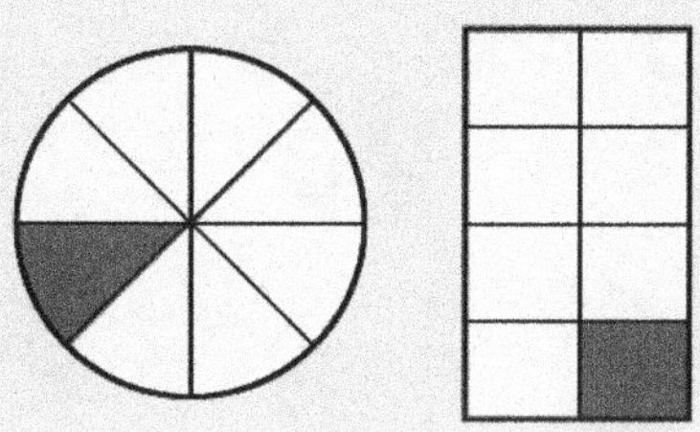	**1/8**	**One eighth**
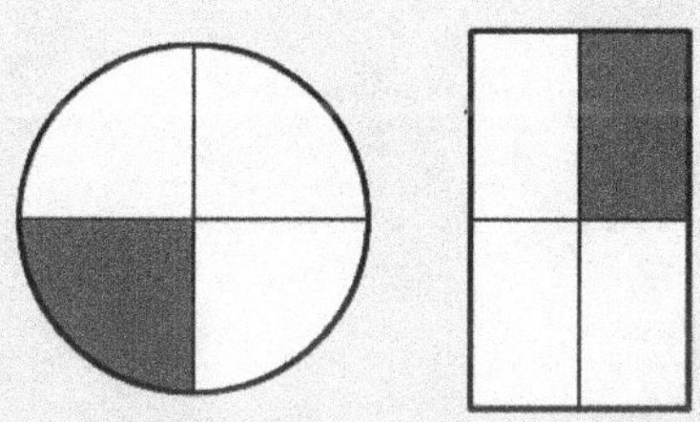	**1/4**	**One fourth**
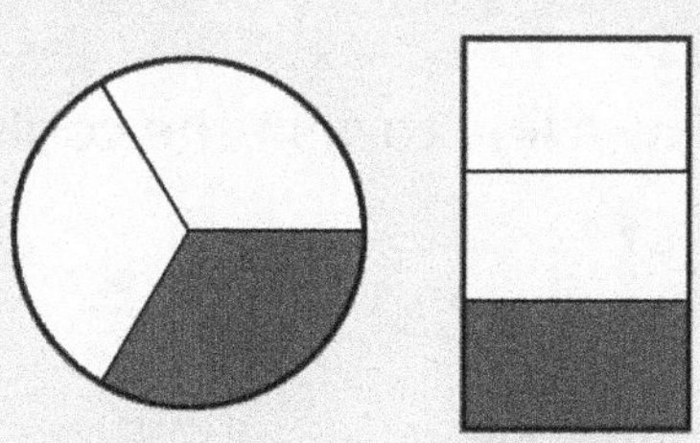	**1/3**	**One third**
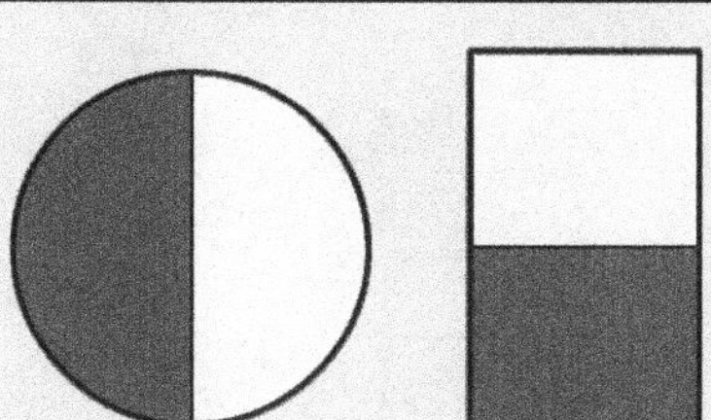	**1/2**	**One half**
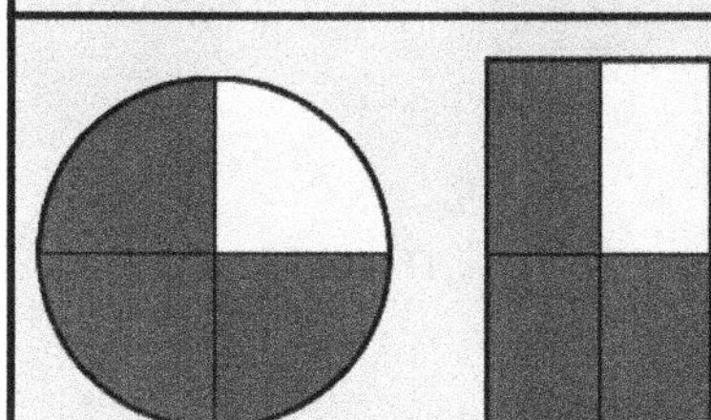	**3/4**	**Three fourths**

Fractions

Level 1

1) Two kids are going to play with 10 marbles, so they empty the bag and split them equally. How many marbles does each kid get?
Hint: Try acting this out with real objects.

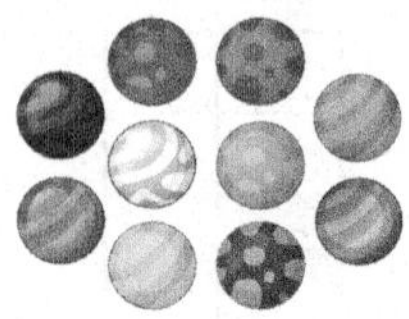

2) I have 12 crackers to pass out to 3 kids. Draw lines from each one to show how I can give them fairly.

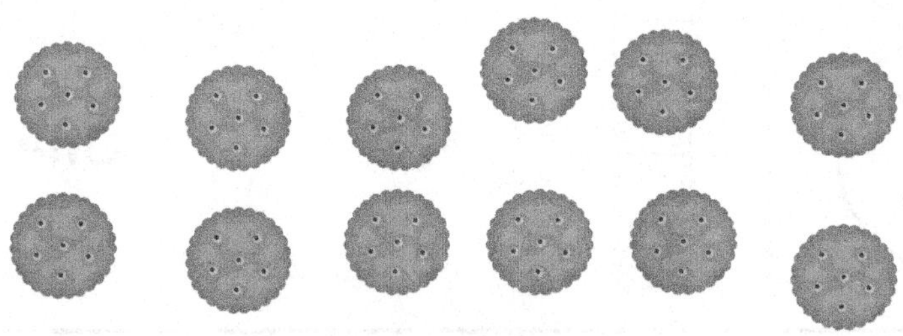

3) If one cookie was going to be shared by two people, draw a line to split the cookie fairly.

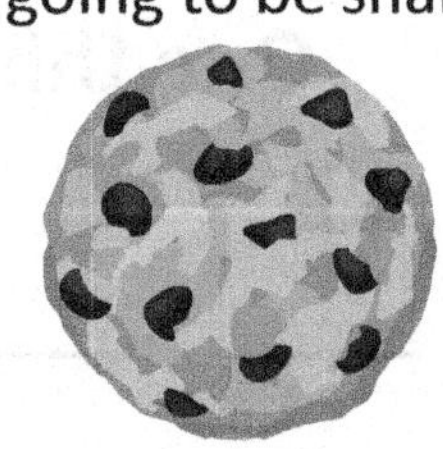

4) Draw lines to split this pie fairly between four people.

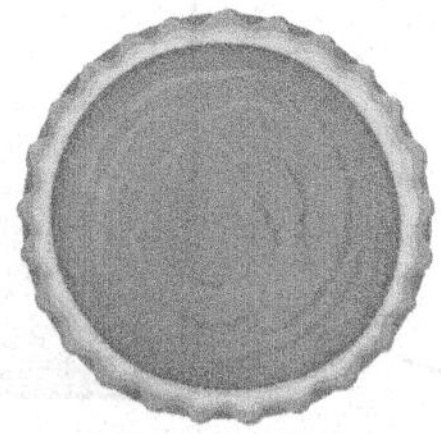

5) My sister split a cookie and said "There are two pieces so that means I'm giving you **half**." Why is she wrong?

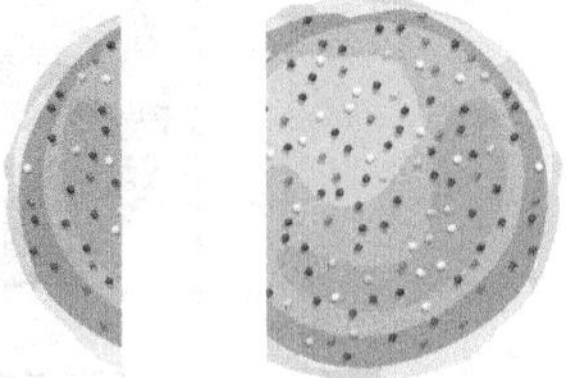

6) Four people split a whole pie. Shade the amount of pie that **two** of them ate.

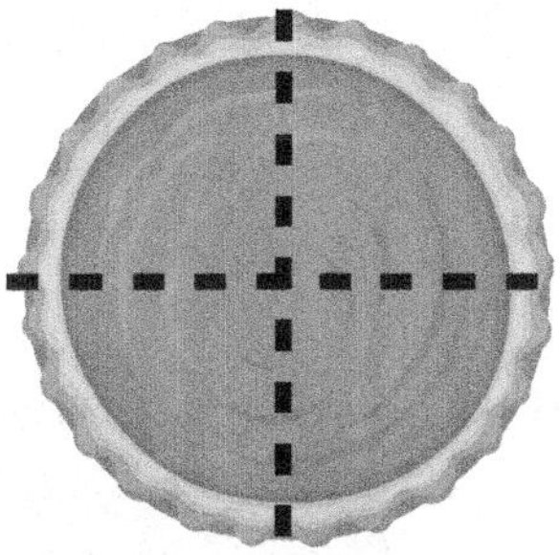

7) What fraction of the pizza is missing?

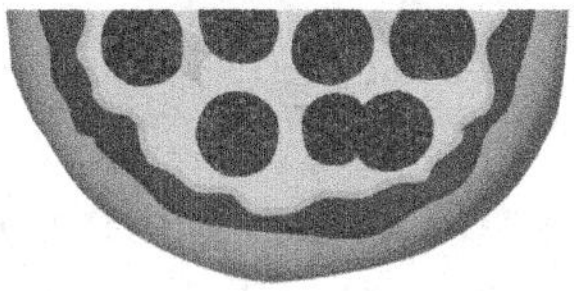

8) One chocolate bar needs to be split fairly between 4 people. Draw lines in the bar to show how to split it fairly. What fraction of the chocolate bar will each person receive?

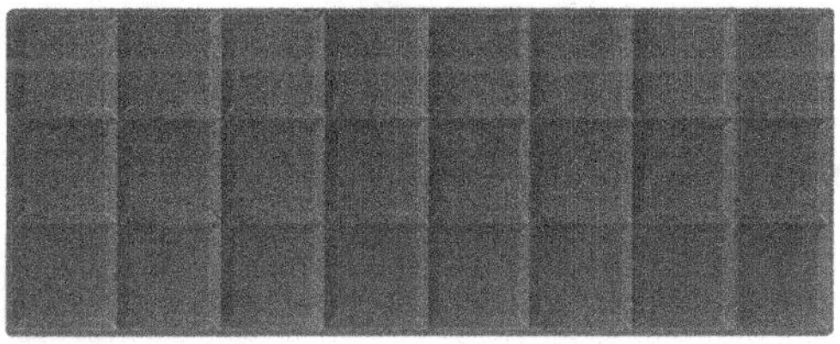

9) Jack and Jill each have 2½ crackers. If they both put their crackers on a plate, how many total crackers would there be?

10) How many cookies would you add to the right plate to make them fair?

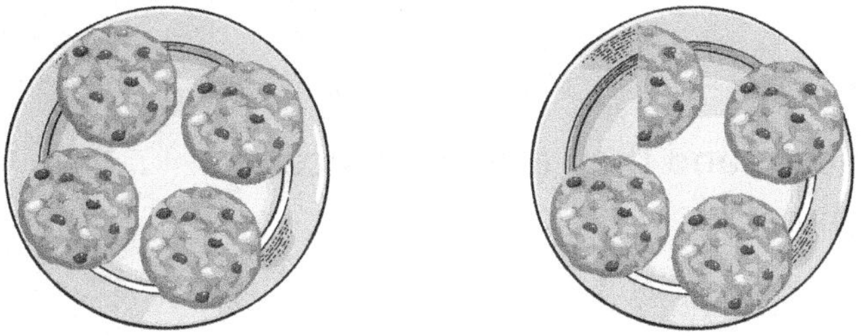

Fractions

Level 2

1) 5 chocolate bars need to be split fairly between 2 people. How many chocolate bars will each person receive?

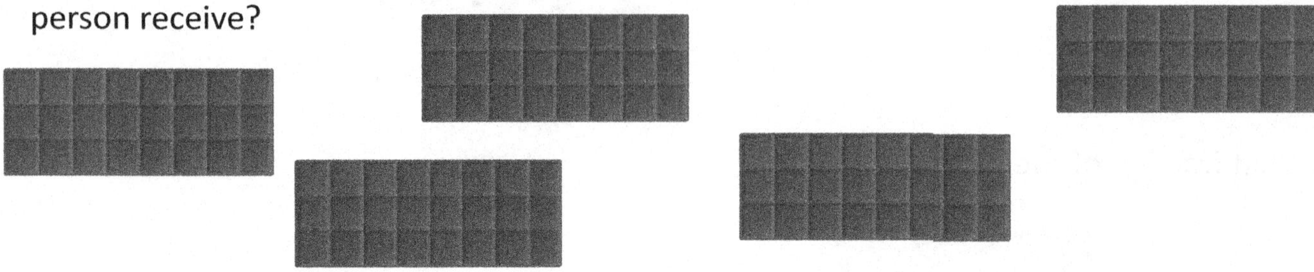

2) Two pizzas were split between four people. Draw lines to divide them evenly, then shade in the amount of pizza that **one** person ate. How much pizza is this?

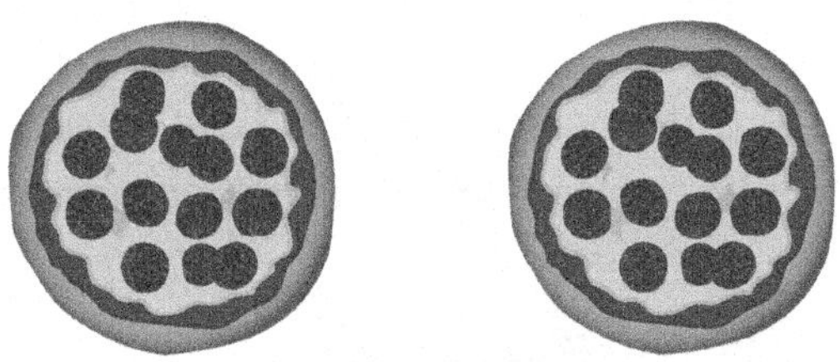

3) How many cookies would you add to the plate on the right so they have the same amount?

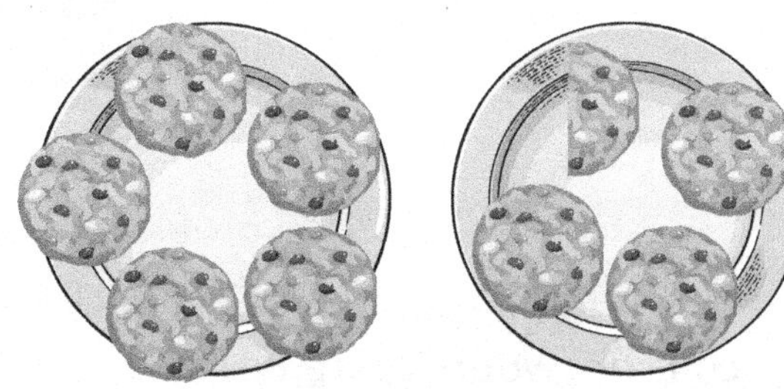

4) Liam, Linda and Laura each have 2½ crackers. How many total crackers would they have if they combined them?

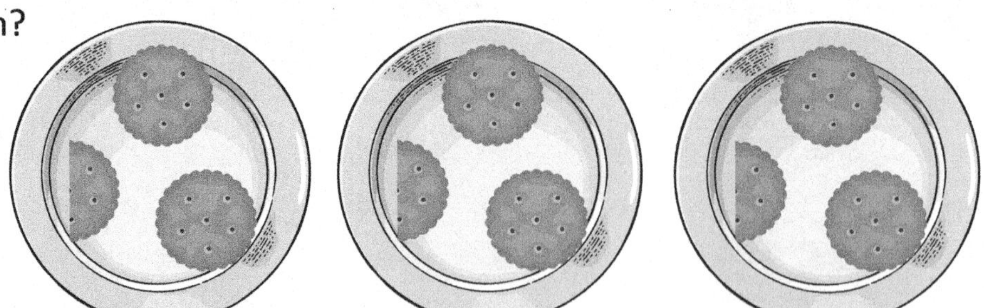

5) Evelyn cut a pie into 8 equal pieces and then ate one piece. How much of the pie did Evelyn eat?

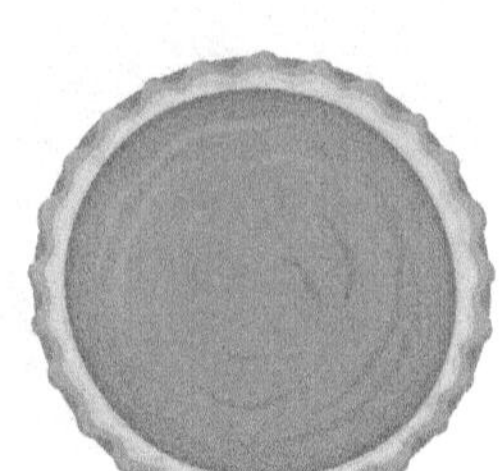

6) Which slice of pizza is bigger: ¼ of the pizza, or ⅛ of the pizza? Divide and shade the pizzas below to check your answer.

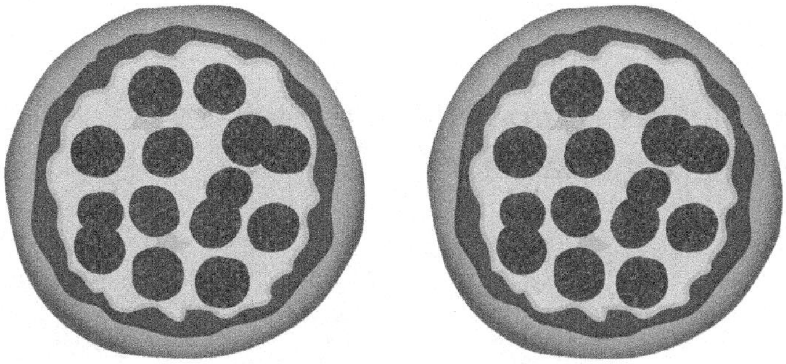

7) I saw this pizza and said "½ of it has pepperoni," but my brother said "2/4 of it has pepperoni." Who is right?

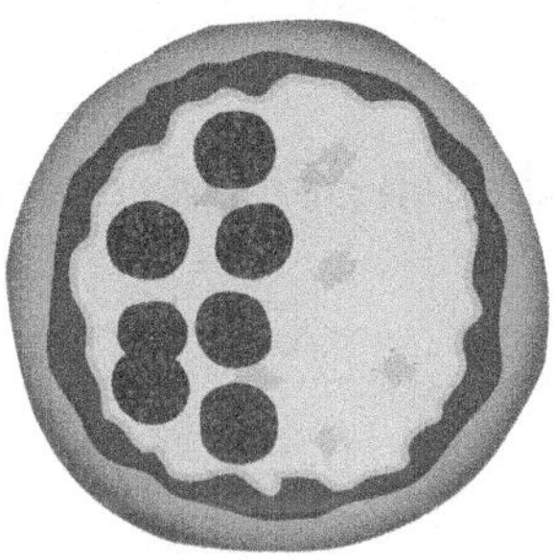

8) Which is bigger, ½ or ⅛? Draw a picture to show why.

9) What fraction of the circle is shaded?

10) Jared gave ½ of his red marbles and ½ of his blue marbles to his sister, but when she counted them, she got confused because she had **12** red marbles and **20** blue marbles.

"½ and ½ are the same fraction, so I should have the same amount for each color."

What does she not understand?

Fractions

Level 3

1) How many cookies would you add to the left side of the scale to make it balance?

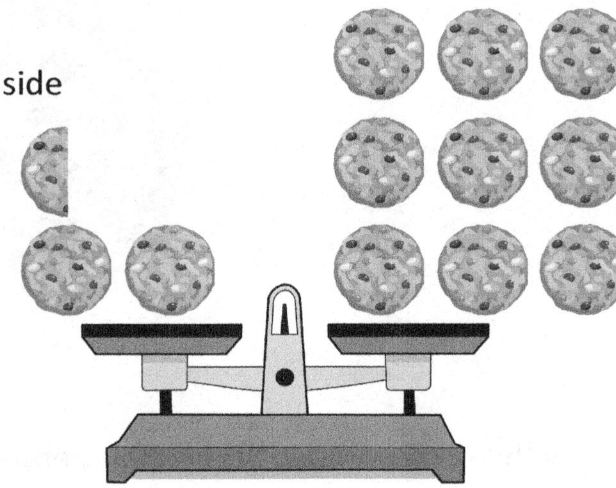

2) 10 people bought 5 pizzas. What fraction of a pizza will each person receive?

3) If you needed to share ¼ of a pie with a friend, how much pie would each of you receive?

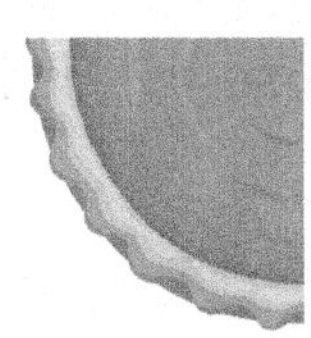

4) Evelyn split a pie into 8 equal pieces and then ate 3 pieces. How much of the pie is left?

5) Eric ate ½ of a pie in the morning and then ¼ of a pie at dinner. His dog Gizmo ate a whole pie. Eric was mad because Gizmo had more pie than he did. How much more pie did Gizmo eat compared to Eric?

6) Which is bigger, $\frac{5}{8}$ or $\frac{9}{10}$? Draw a picture to show why.

7) What is $\frac{2}{4} + \frac{1}{4}$?

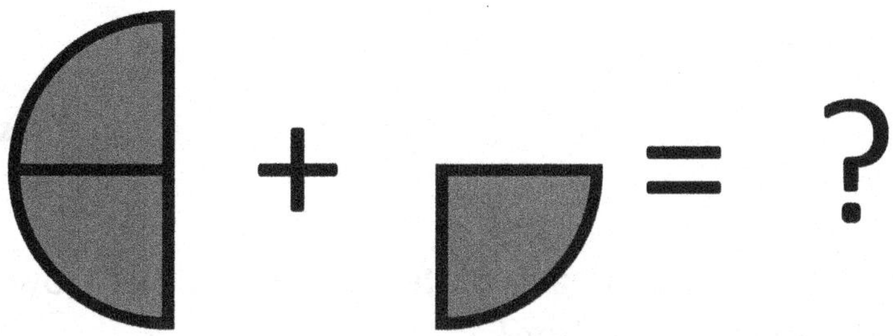

8) What is $\frac{1}{8} \times 4$? *Hint: Think about pizza slices.*

9) What is $\frac{1}{2} \div 2$? *Hint: Think about dividing half of a pizza.*

10) My friend challenged me to eat $\frac{3}{2}$ of a pizza. Draw a picture to show how much that would be.

Fractions

Genius Level

1) Each time a frog jumps, it travels 1¼ feet. How many jumps would it need to travel 10 feet? *Hint: Draw a picture.*

2) There were 32 plates at a table, and each plate had ⅛ of a pie. How many whole pies would there be if we put all the pieces together?

3) If 2 cookies are going to be shared with 16 people, what fraction of a cookie would each person receive?

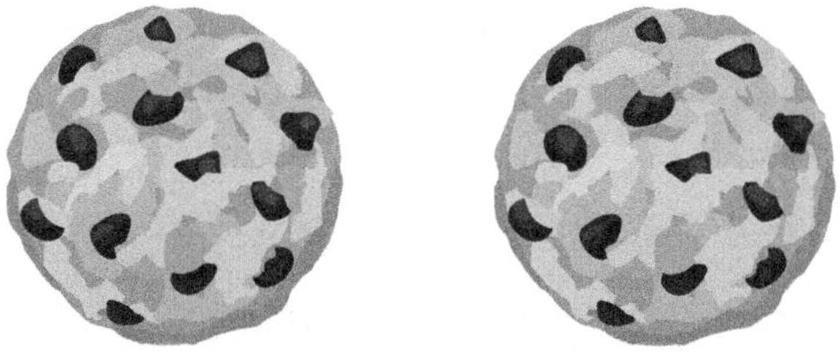

4) How many cookies would you add to the right side of the scale to make it balance?

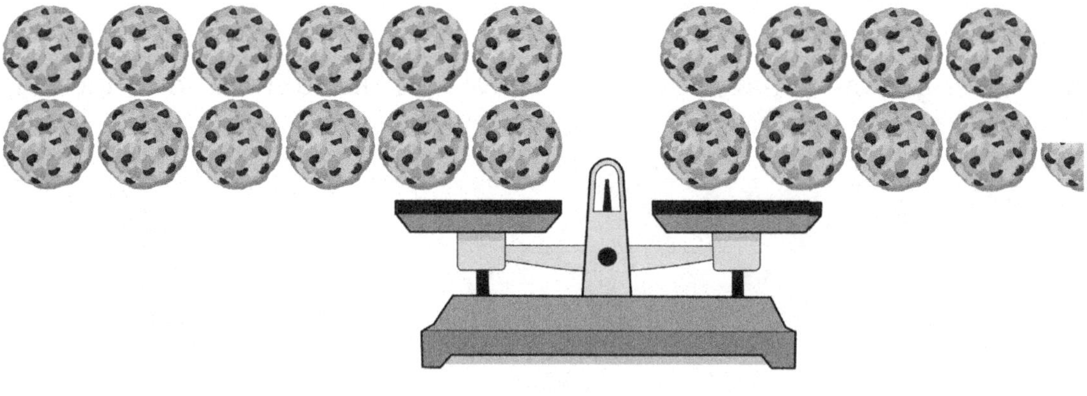

5) Peggy found a bag of money. She gave ½ to charity and ¼ to her brother. If she had $5 left, how much money was in the bag when she found it?

6) Gabe said he was 5¾ years old. How old will he be on his next birthday?

7) Which is bigger, ⅗ or ⅔?

8) Rachel was given a piece of pie, then she ate ½ of that piece. What she had left was ⅛ of the whole pie. How much of the pie did she start with?

9) ⅞ of a mystery number is 112. What is the mystery number?

10) On Monday, Peter picked ½ of the carrots in his garden. On Tuesday, he picked ½ of what was left, then he ate one more carrot as a snack. That night there were 5 carrots left in his garden. How many carrots were in the garden before Peter started picking them?

Hint: Work backwards!

19

Percents

Percents are just another way to talk about parts of things. Look at this half pie. We can say there is "½ left" or there is "50% left." They both mean the same thing.

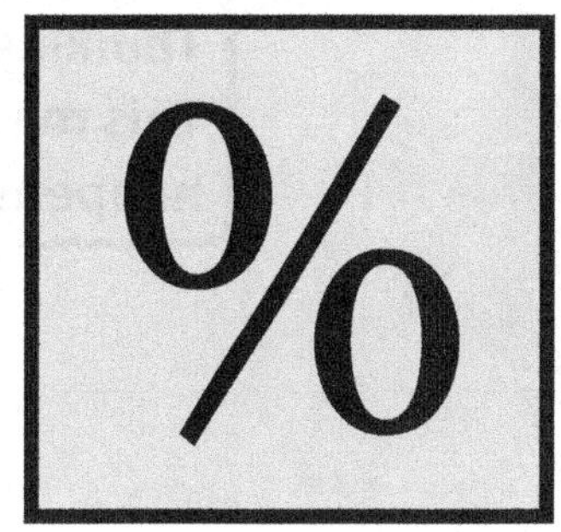
%

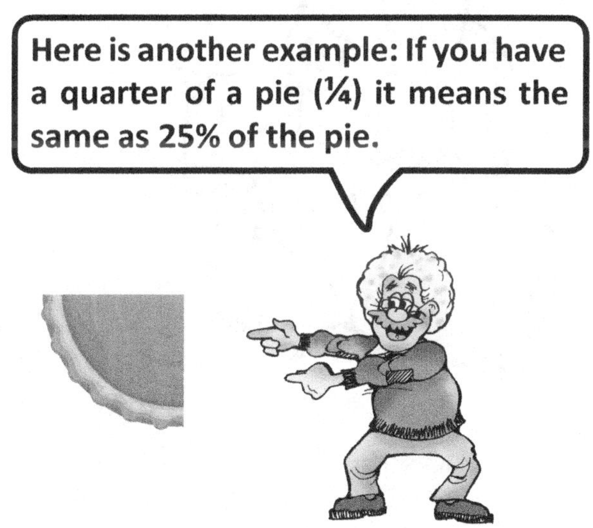
Here is another example: If you have a quarter of a pie (¼) it means the same as 25% of the pie.

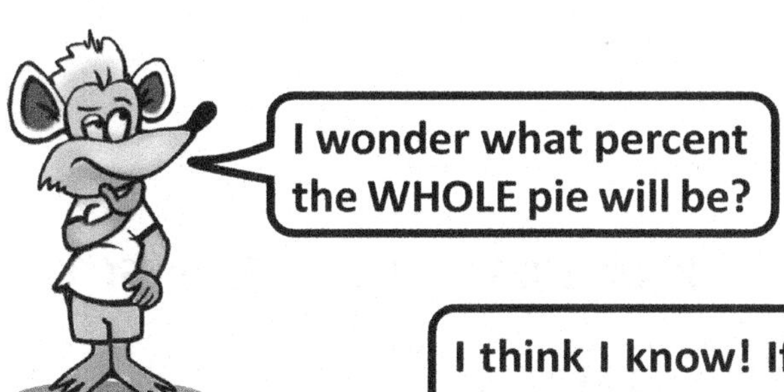
I wonder what percent the WHOLE pie will be?

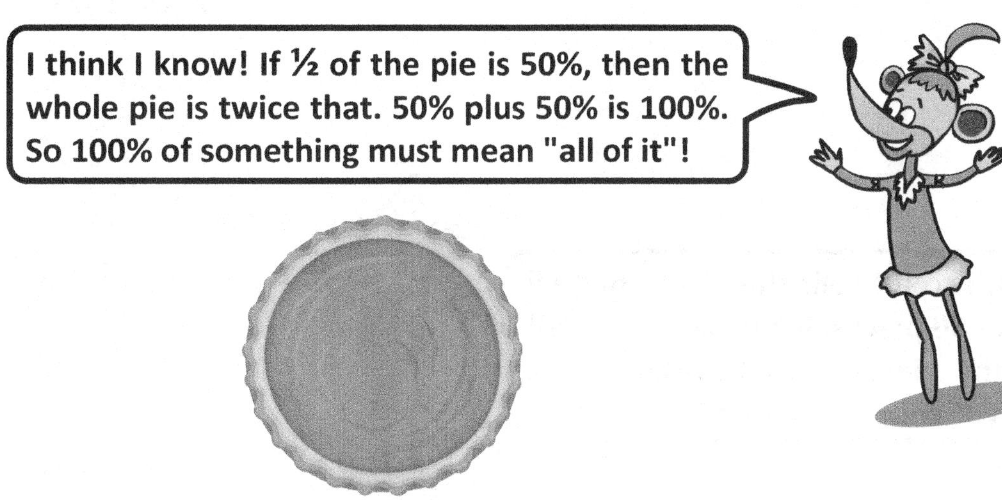
I think I know! If ½ of the pie is 50%, then the whole pie is twice that. 50% plus 50% is 100%. So 100% of something must mean "all of it"!

What "Percent" Means

That's right! When we say "percent," we mean "how many of the hundred pieces."
I see percents a lot when I work with money. If a soda costs $1 and there is a 5% sales tax, I wonder how much money that 5% tax will be.
$1
+ 5% tax
Let's put a dollar through the Percent Machine to see what happens!
% Machine
Your machine split the dollar into 100 pennies! So each penny is 1% of a dollar. That means 5% is 5 pennies!
100 pennies is 100 cents.
50 pennies is 50 cents.
1 penny is one cent.
That sounds a lot like "percent!"
That is because the word "cent" is Latin for "hundred." 100 cents means 100 pennies which is the same as 1 dollar.

Percents vs. Fractions

50% being ½ makes sense to me, but I don't understand more complicated numbers. How is something like 67% related to fractions?

Remember that percents are just another math language! Kind of like "el gato" is a Spanish way of saying "the cat."

Remember how the bottom part of a fraction (denominator) means how many "equal parts" a number was split into?

$$\frac{1}{2}$$

1 ← How many pieces you have

— ← (divided by)

2 ← how many **equal parts**

Oooh, "equal parts" sounds familiar. If percents split something into a HUNDRED equal parts, that means 67% is like a fraction with "100" on the bottom! It's like we split the number 1 into 100 pieces and took 67 of them.

$$67\% = \frac{67}{100}$$

A double-number line is one way to see the relationship between percents and fractions.

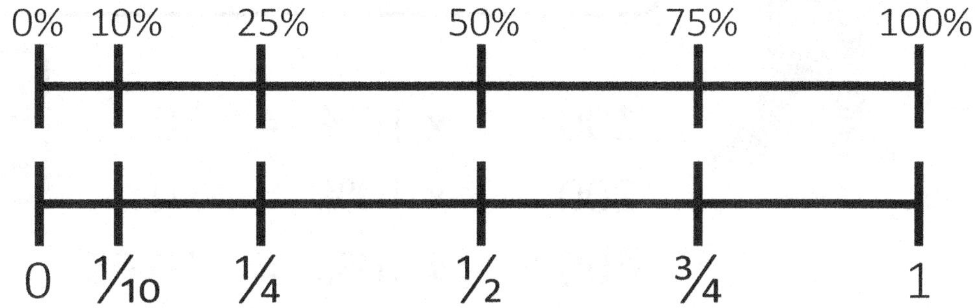

I know we *could* have put more fractions and percents on that line, but those seem like the most important numbers to remember.

If 10% means the same as 1/10, that seems like an easy percent to find! We know that taking a tenth of a number is the same as dividing it by 10. So 10% of 50 is 5!

$$50 \div 10$$

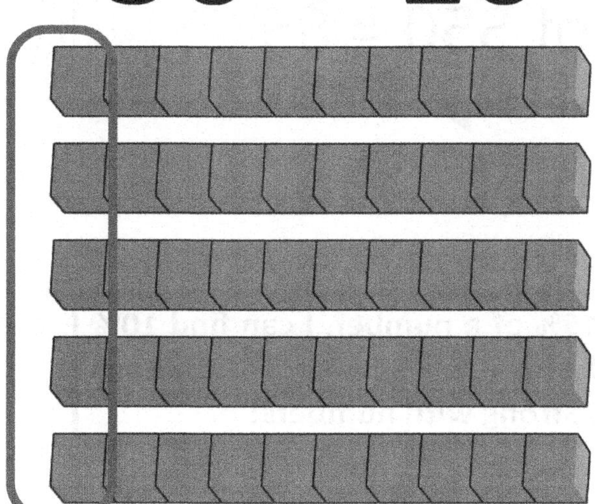

The Power of 10%

You have discovered an important idea to help find some percents quickly! Look what happens when we take 10% of these numbers...

100	x 10%	=	10
200	x 10%	=	20
100.5	x 10%	=	10.05
\$12.56	x 10%	=	\$1.256

Those dots are called Decimals. We will learn about those in the next chapter!

That's great I can divide by 10 to find 10%, but I wish this helped with the other percents.

I think it *can* help! If I need to find 20% of a number, I can find 10% quickly and then double it!

20% of \$50?

10% of \$50 = \$5

20% is \$5 x 2

And to find 15% of a number, I can find 10% and add half of it (\$5 + \$2.50 = \$7.50). I am really getting strong with numbers!

Percent Change

You have come a long way so far! Percents can also describe how a number *changes*. Look what happens when a 10-inch worm grows by 50%.

A 10-inch worm grew by 50%. How long is it now?

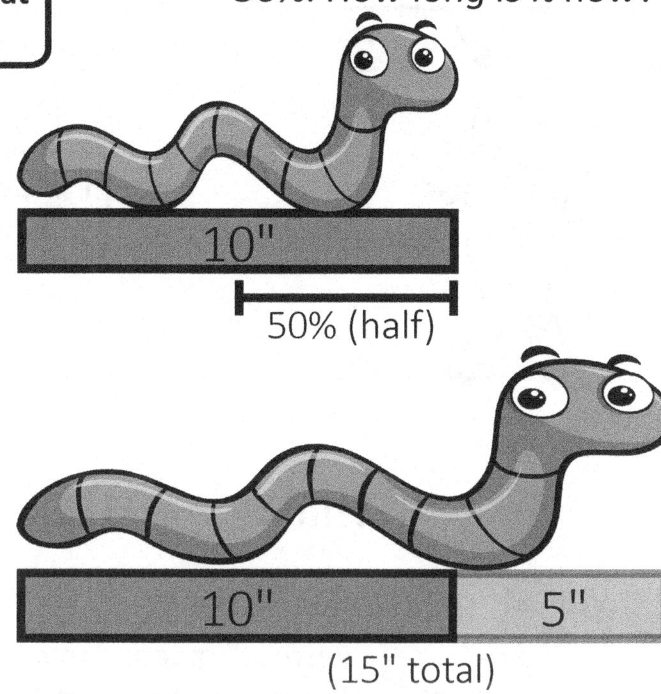

A worm grew from 10 inches to 20 inches. What percent did it grow by?

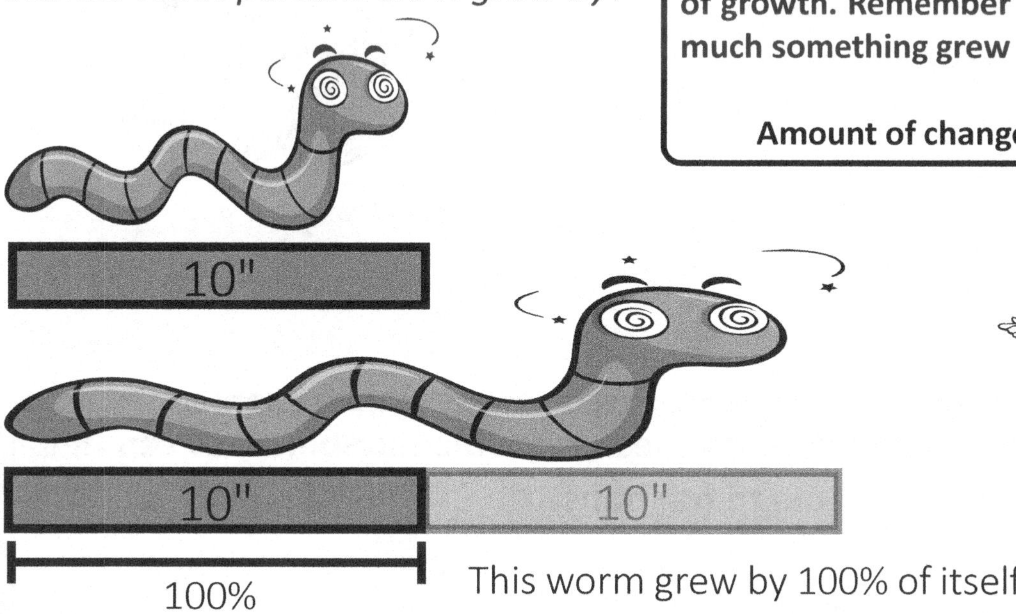

We can also work backwards to find the percent of growth. Remember this shortcut to see how much something grew or shrunk:

Amount of change ÷ Starting Amount

$$\frac{\textbf{Amount of change}}{\textbf{Starting Amount}}$$

Final Strategies

Use Benchmarks

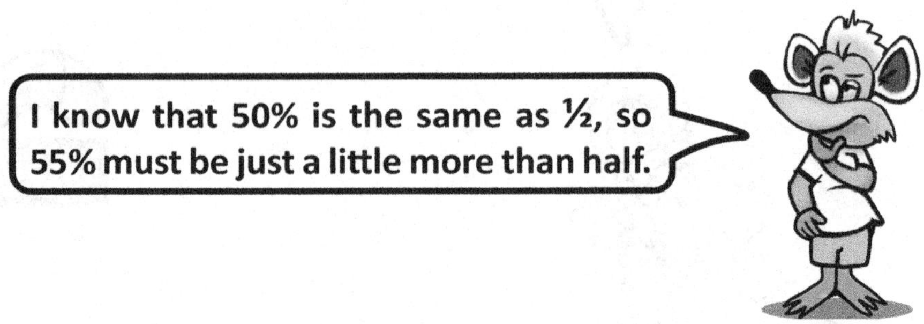

Think __/100

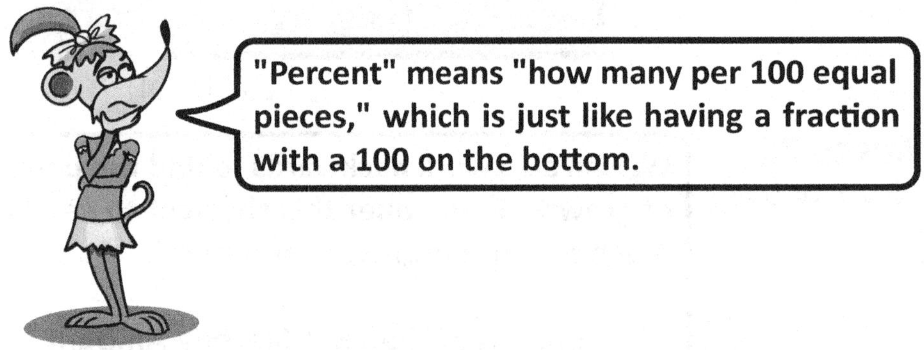

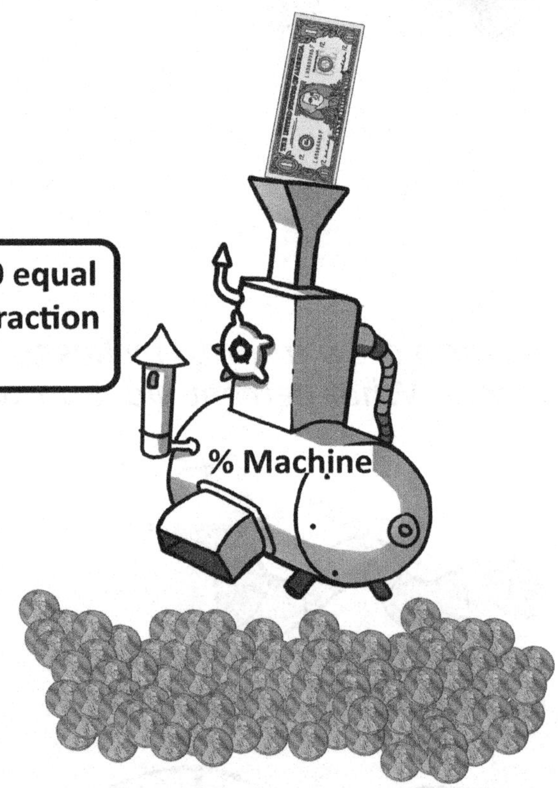

Remember, drawing rectangles or double number lines can help you compare percents to fractions.

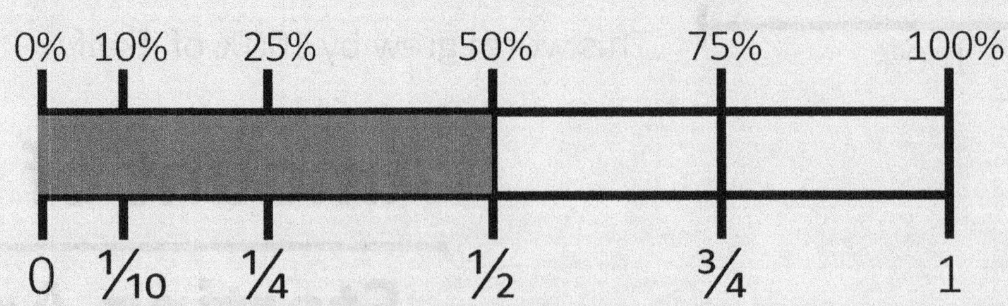

Percent Benchmarks

Here are some important percent benchmarks to remember:

Percent	*Fraction*	*Visuals*
10%	**1/10**	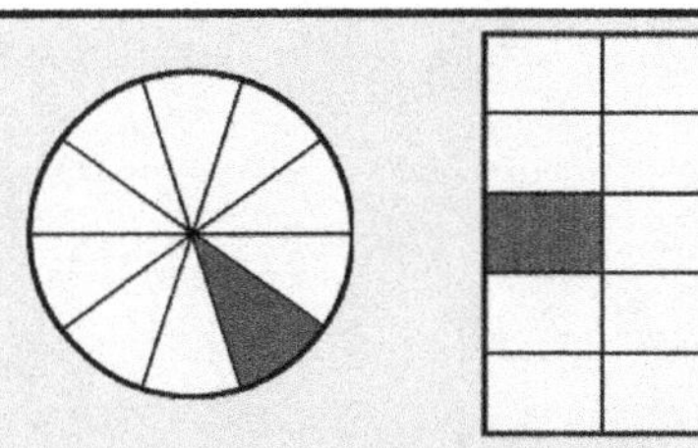
20%	**1/5**	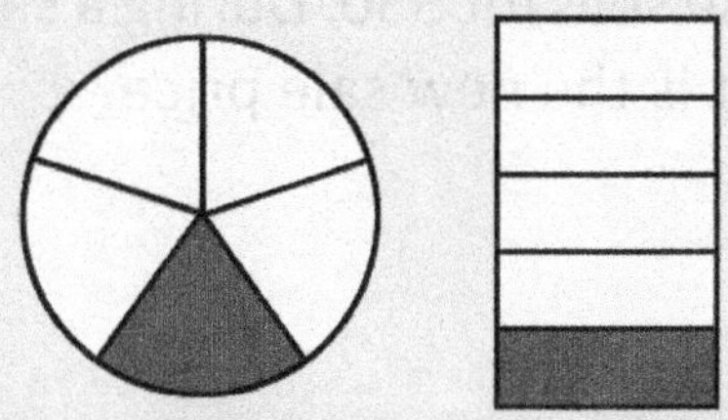
25%	**1/4**	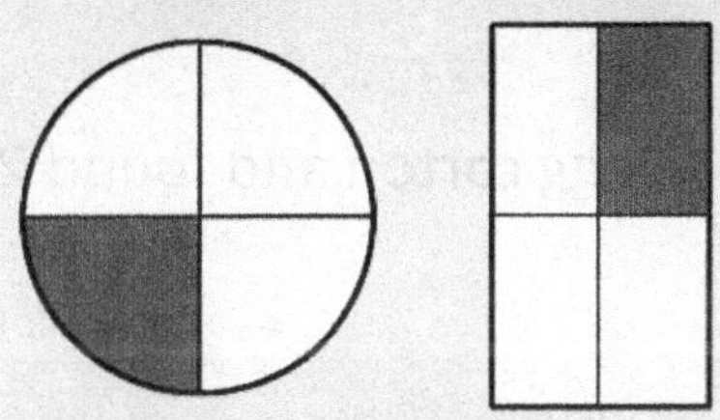
33.33% *(This repeats forever)*	**1/3**	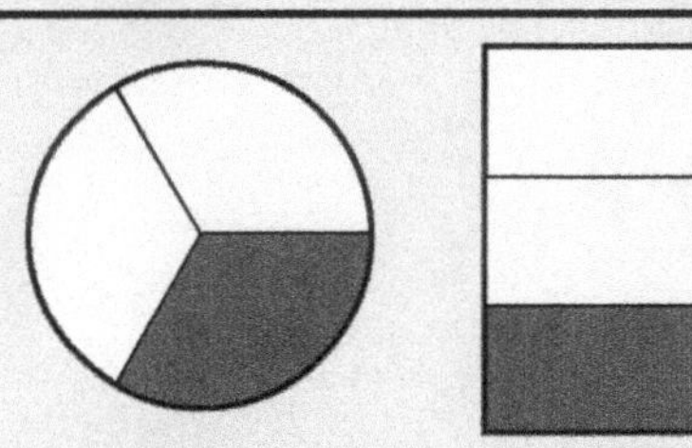
50%	**1/2**	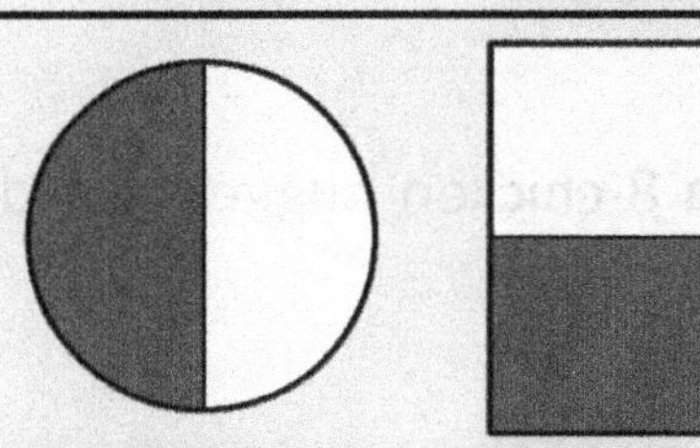
75%	**3/4**	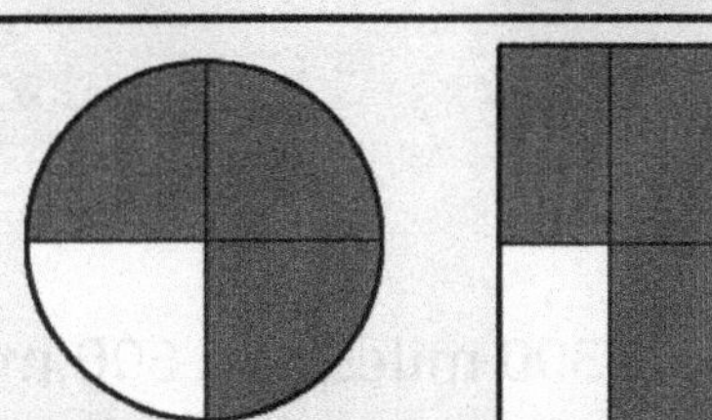
100%	**1**	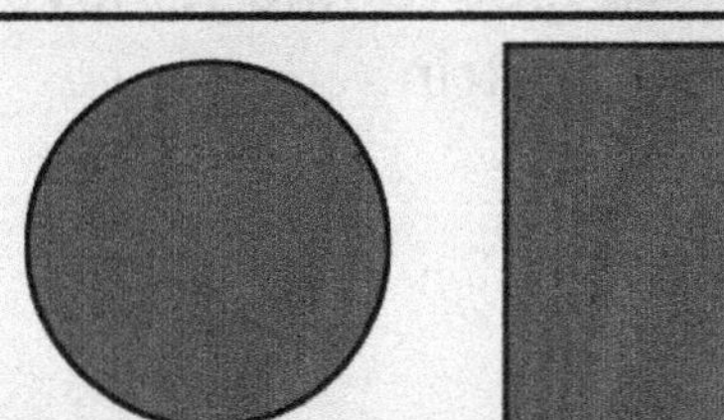

Percents

Level 1

1) Mark decided to give 50% of his $8 allowance to charity. How much will he give to charity?

2) A box of cheese sells for $30. During a sale, the price will be 50% cheaper. What is the new sale price?

3) Noah opened an egg carton and found 25% of the 12 eggs were broken. How many eggs were broken?

4) Seth was given 8 chicken nuggets for dinner and he ate 100% of them. How many chicken nuggets did Seth eat?

5) If a family travels 300 miles of a 600 mile trip, they have traveled 50% of the trip. If a slug needs to crawl 40 feet to find its favorite food and has gone 10 feet so far, the slug has finished ____% of its trip.

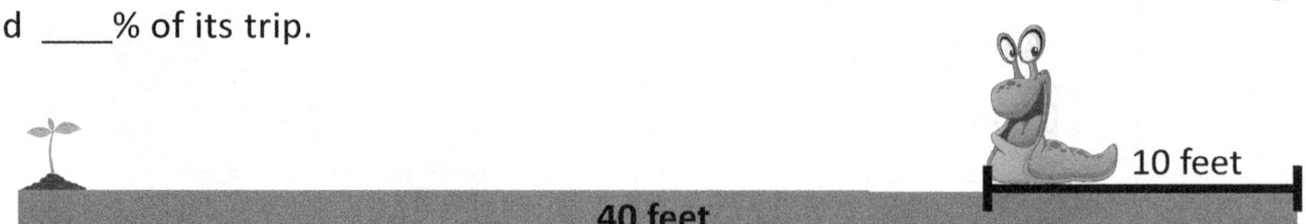

6) The number 30 is 50% of 60. The number 75 is 50% of what number?

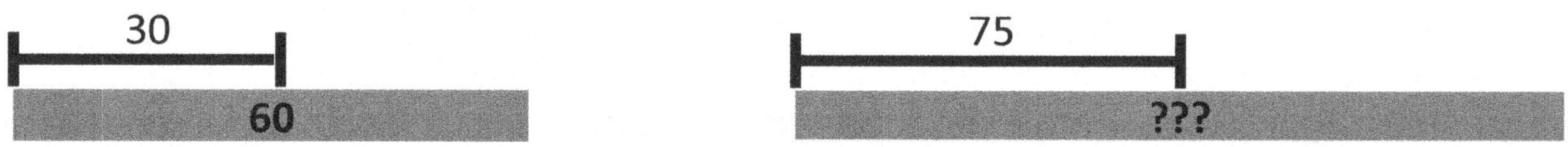

7) A mouse ate 10% of a 10 pound block of cheese. How much cheese did the mouse eat?

8) 10% of $80 is $8. What is 5% of $80?

9) Pete spent 25% of his money. If he started out with $800, how much money does he have left?

10) 10% of $120 is $12. What is 20% of $120?

Percents

Level 2

1) A snowboard cost $100. A clerk said the total was $107 including tax. What percent tax was there?

2) A family needs to decide how much to tip a waiter after their meal. The cost of the meal was $80 and they want to leave a 15% tip. How much tip should they leave?

3) What is 75% of 8?

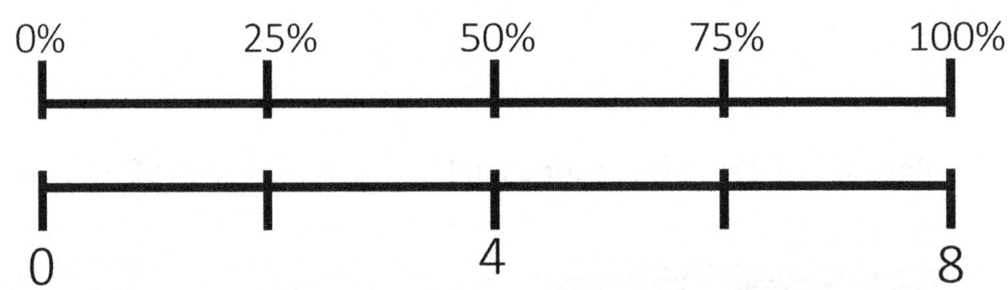

4) What is 2% of one dollar?

5) If a 50-pound boy can lift 10 pounds, he can lift ____% of his weight.

6) Aria's allowance is the same as Gabe's allowance. Aria's allowance is ____% of Gabe's allowance.

7) If a 10 pound dog can lift 8 pounds, it can lift 80% of its weight. If an 80 pound chimpanzee can lift 20 pounds, it can lift 25% of its weight. If a 100 pound child can lift 32 pounds, he can lift ____% of his weight.

8) What is 100% of 86,551?

9) Stu wanted to buy a bike that cost $200, but the next week the price increased to $300. By what percent did the price increase?

10) If Pete the Dragon started out with $1000, how much money will he have left if he spends 90% of it?

Percents

Level 3

1) If \$200 is sent through the percent machine, the \$200 is broken into 100 equal pieces. How much money is each piece worth?

2) ½ = 50%
   ¼ = 25%
   ⅒ = 10%

   ⅕ = ____%

3) Each floor in a building is about 10 feet tall. In a 50-foot building, what percentage of the building's height is one floor?

4) What is a 20% tip for a \$150 meal?

5) 100% of 52 is 52. ____% of 52 = 104?

6) Joseph's allowance was $10 per week. When he turned 12 years old, his parents said they would give him a 50% raise. His raise is $5 (because 50% x $10 = $5), so his new allowance is $15 per week. What would his new allowance be if his parents gave him a 10% raise instead?

7) 1% of $175 is $1.75
1% of $560 is $5.60
1% of $75 is 75 cents
1% of $145 is $1.45

What is 2% of $725?

8) A magic lizard was 50" long and shrunk to 25". By what percent did it shrink?

9) If Beth has $500 in a savings account that earns 5% interest at the end of each year, how much interest will Beth earn in the one year?

10) A credit card charges a yearly interest of 20%. If you buy a $100 toy with a credit card and then wait a year to pay, you will end up paying $100 for the toy AND $20 in interest.

Rick bought a $500 computer with a 20% interest credit card and waited a year to pay back the $500. How much interest did Rick have to pay?

Percents

Genius Level

1) If a 100" snake grows by 50% each year, how long will the snake be in 2 years?

2) Noah opened an egg carton and found that 9 of the 12 eggs were broken. What percent of the eggs were **NOT** broken?

3) My piggy bank started with $50 in it. I spent some money and now there is $5 left. What percent of the original $50 did I spend?

4) ½ = 50%
   ¼ = 25%

   ⅘ = ____%

5) My dog's weight decreased by 10% and now he weighs 45 pounds. How much did he weigh before his diet?

6) Gabriel and his friend Nicky each have 2 rats. Gabriel said that he has 100% of the rats that Nicky has.

Gabriel then received 2 more rats for Christmas, and said: I no longer have 100% of the rats you have, I now have ____% of the rats you have.

7) My red fish is 12 inches long and my blue fish is 4 inches long. What percent length is my red fish compared to the blue fish?

8) ½ = 50%
¼ = 25%

$^{19}/_{20}$ = ____%

9) ½ and ½% are very different numbers.

½ of 1000 is 500
1% of 1000 is 10, so
½% of 1000 is 5

What is ½% of $100?

10) Joseph's allowance was $10 per week. What would his new allowance be if his parents gave him a 2½% raise?

20

Decimals

Percents, fractions, decimals—they are all just a different math language to describe the same amount.

25%	*Percent*
$\frac{25}{100}$	*Fraction*
.25	*Decimal*

We can see decimals when numbers move to the RIGHT on a Place Value mat (the opposite direction we have been used to so far).

Tens	Ones	Tenths	Hundredths	Thousandths
10	**1** •	**$\frac{1}{10}$**	**$\frac{1}{100}$**	**$\frac{1}{1000}$**

When we move left, the Grow Monster *multiplies* each column by 10...

...but when we move right, the Shrink Monster *divides* each column by 10.

Decimals Expanded

Tens	Ones	Tenths	Hundredths	Thousandths
10	1 ●	$\frac{1}{10}$	$\frac{1}{100}$	$\frac{1}{1000}$

34.25

Tens	Ones	Tenths	Hundredths
3	4 ●	2	5

$$30 + 4 + \frac{2}{10} + \frac{5}{100}$$

From Decimals to Fractions

I'm starting to understand, but I need it explained a different way to make sure.

Maybe it will help to look at this decimal:

.5

	Tenths	Hundredths
•	5	

This is in the "tenths" column, so we would say it is "5 tenths." Hey, that sounds like a fraction! 5 tenths...5⁄10

Now I see the connection between decimals and fractions—they are just a different way of saying or writing the same amount!

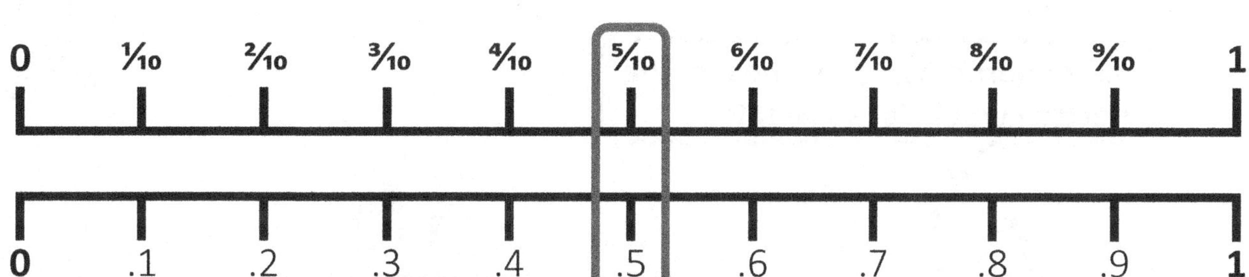

It helps to think of the double number line from before. I see the pattern now—It looks like the decimal is just the top part of a fraction!

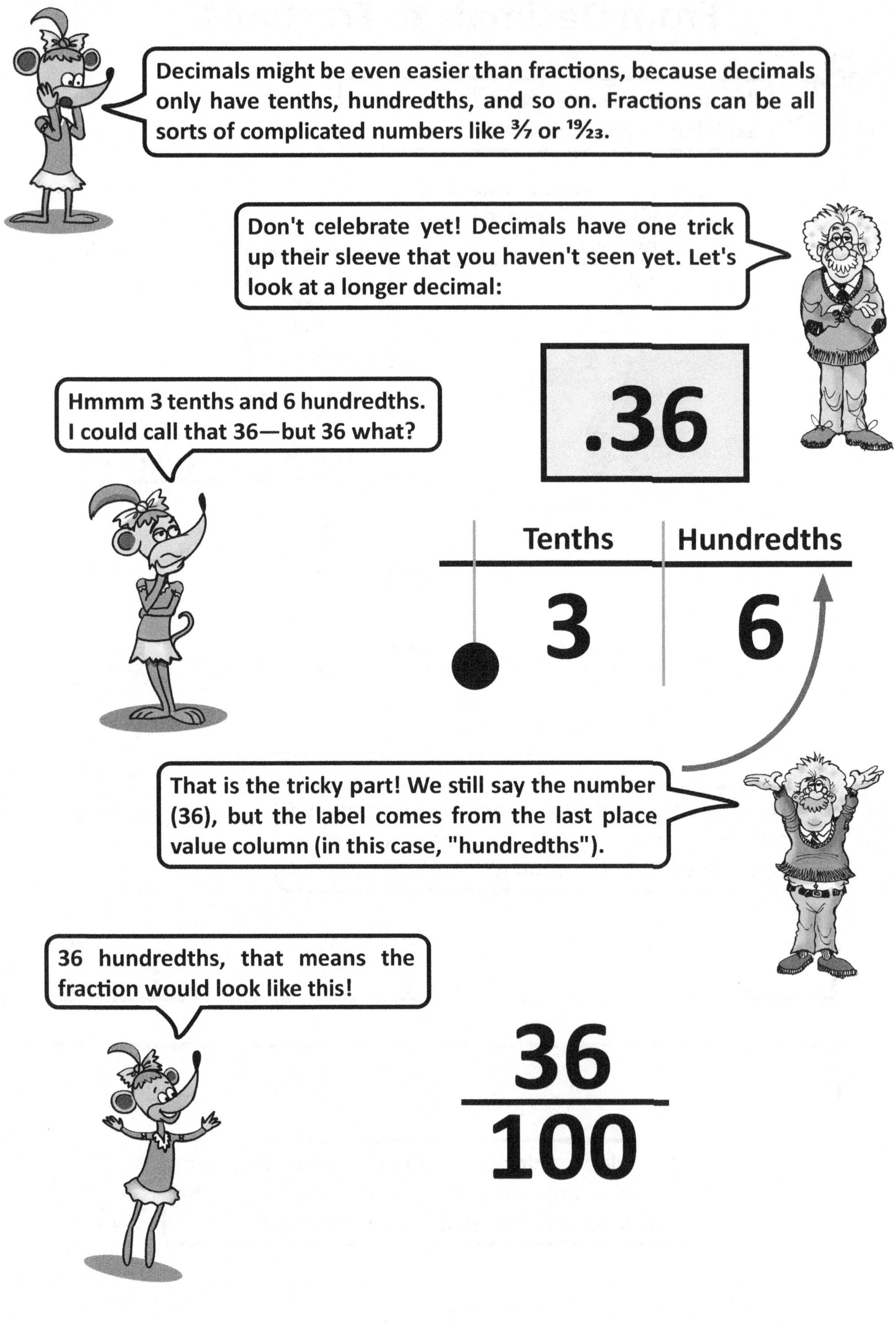
Decimals might be even easier than fractions, because decimals only have tenths, hundredths, and so on. Fractions can be all sorts of complicated numbers like 3/7 or 19/23.
Don't celebrate yet! Decimals have one trick up their sleeve that you haven't seen yet. Let's look at a longer decimal:
.36
Hmmm 3 tenths and 6 hundredths. I could call that 36—but 36 what?
Tenths
Hundredths
. 3 6
That is the tricky part! We still say the number (36), but the label comes from the last place value column (in this case, "hundredths").
36 hundredths, that means the fraction would look like this!
36/100

0.1	**1 tenth**	**$\frac{1}{10}$**
0.15	**15 hundredths**	**$\frac{15}{100}$**
0.243	**243 thousandths**	**$\frac{243}{1000}$**

0.1

Tens	Ones	Tenths	Hundredths
	0 ●	**1**	

Last week I saw a decimal that really confused me:

.04

Let's put it on the place value mat to see what happens...

Tens	Ones	Tenths	Hundredths
	•	0	4

Oh, that makes much more sense now! There are no tenths, so I don't need to worry about that column.

All I have are "4 hundredths." That's like the fraction $\frac{4}{100}$!

See page 338 for a blank Place Value Mat with decimals!

Reverse: From Fractions to Decimals

I know how to turn a decimal into a fraction, but what about going backwards and turning a fraction into a decimal?

Uh oh—we decimals are picky about which fractions you turn into us. We prefer fractions with a multiple of 10 on the bottom.

/10	/100	/1000	/10000

These kinds of fractions are easy to change into decimals! All I need to do is say their name and remember the place value columns.

$\frac{4}{10}$	4 tenths	.4
$\frac{17}{100}$	17 hundredths	.17
$\frac{4}{100}$	4 hundredths	.04

But what if the fraction doesn't follow the Decimal Man's rules, like this?

$$\frac{4}{7}$$

There is a math way to find the answer, but sometimes a calculator is the best way to turn these kinds of fractions into decimals. The fraction line means "divide," so 4 divided by 7 is:

.57142857142... (and it keeps going on for a while)

Benchmarks

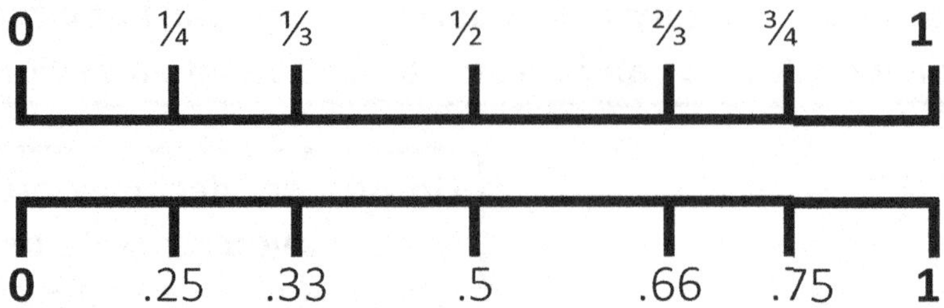

Our double number line can help us remember important benchmarks and use them to estimate fractions or decimals. We see these a lot, so it will help to learn them.

Turning Decimals to Percents

I don't want to leave percents out of the math fun! I know percents are like a special fraction, so how do they relate to decimals?

Remember how percent means "how many per hundred pieces"? Well, that's just another way of saying "hundredth."

Ohh so that means that if a decimal says "hundredth," it will be the same as the percent: .25 = 25%

Decimal	*Words*	*Fraction*	*Percent*
.25	**25 hundredths**	**25/100**	**25%**

What about a decimal like this? I know we say "3 tenths" but I feel like we should be able to make it a percent, too.

.3

To turn ANY decimal into a percent, you can multiply it by 100.

I'm going to multiply .3 by 10 twice so I can watch it gradually turn into a percent.

.3 x 100

.3 x 10 = 3

3 x 10 = 30

.3 = 30%

Adding / Subtracting Decimals

$$\begin{array}{r} 1.14 \\ +\ 0.04 \\ \hline 1.18 \end{array}$$

$$1.14 + 0.04$$

$$1 + \tfrac{1}{10} + \tfrac{4}{100} + \tfrac{4}{100}$$

$$1.18$$

(8 hundredths)

Concrete Tools

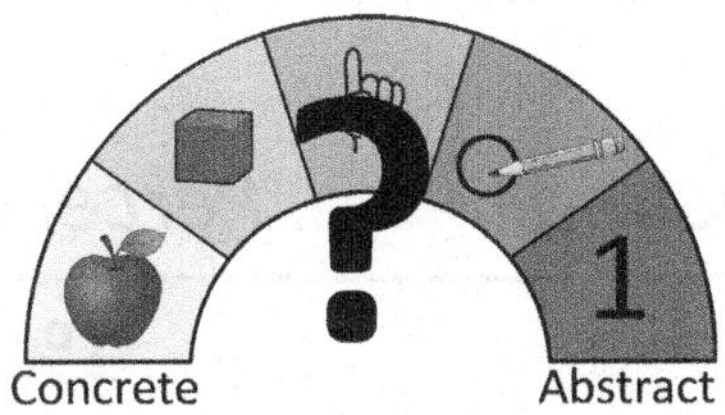

Unfortunately, we don't have a great concrete tool for decimals. They are so small that we can't hold "4 hundredths" the same way we can hold "4."

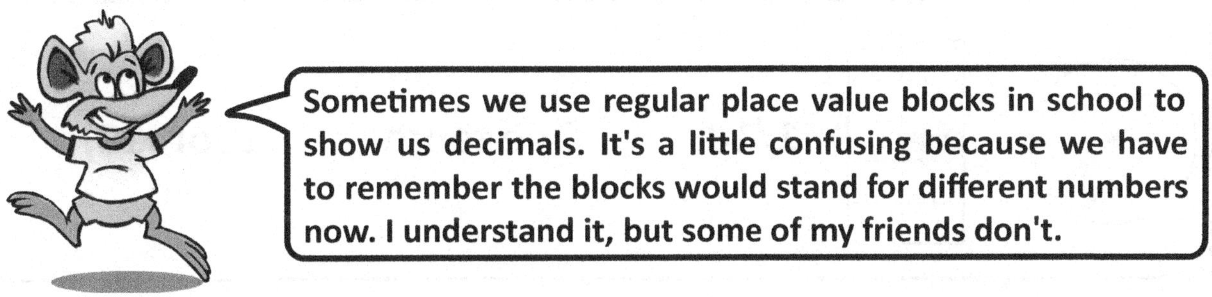

Important Benchmarks (Summarized)

Here are the most important number benchmarks as **fractions, words, decimals,** and **percents.**

Visuals	*Fraction*	*Words*	*Decimal*	*Percent*
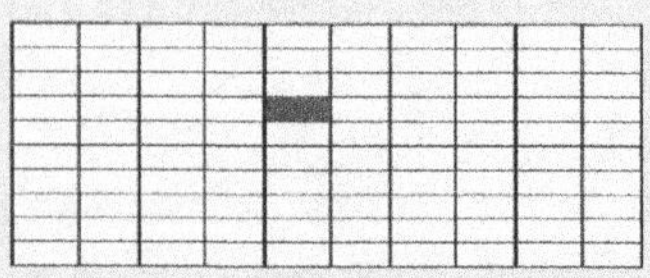	**1/100**	**One hundredth**	**.01**	**1%**
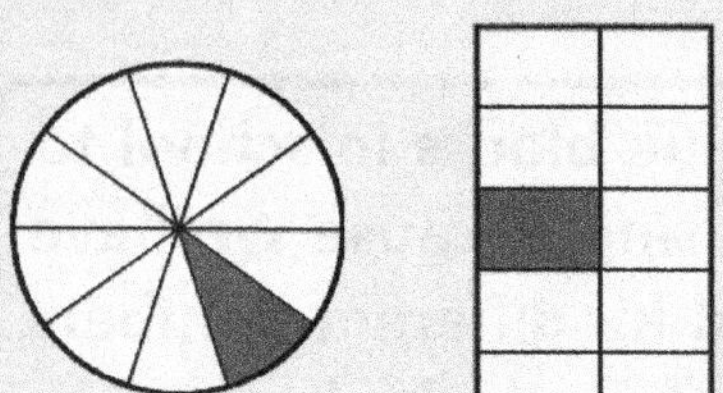	**1/10**	**One tenth**	**.1 (or .10)**	**10%**
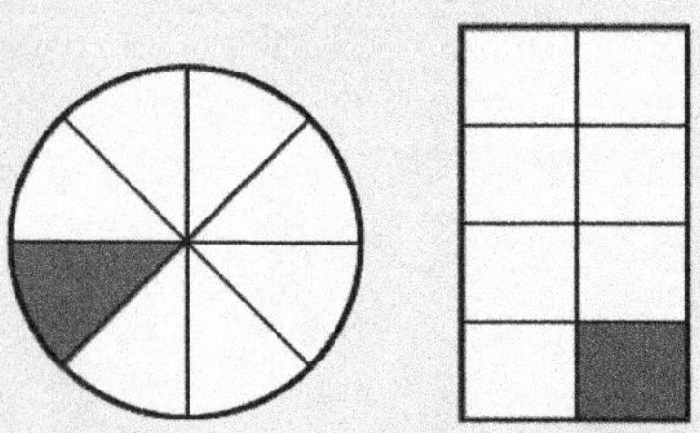	**1/8**	**One eighth**	**.125**	**12.5%**
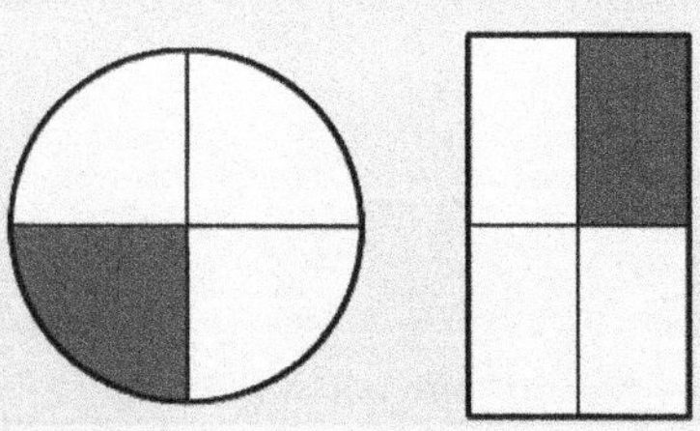	**1/4**	**One fourth**	**.25**	**25%**
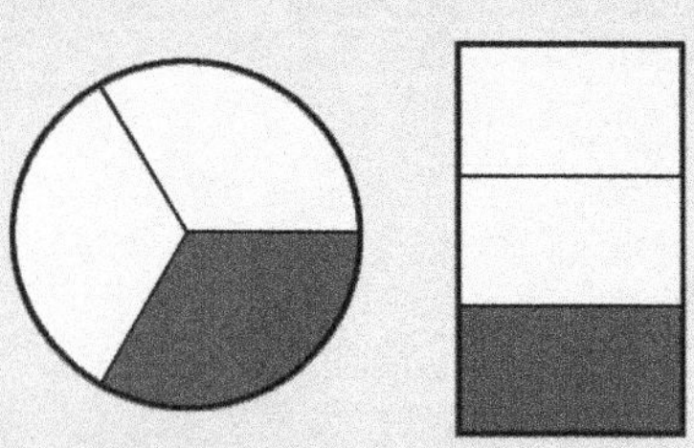	**1/3**	**One third**	**.333333** *(These repeat forever)*	**33.33%**
	**1/2**	**One half**	**.5 (or .50)**	**50%**

Other Comparisons

Here are some other examples (each benchmark doubled).

Visuals	*Fraction*	*Words*	*Decimal*	*Percent*
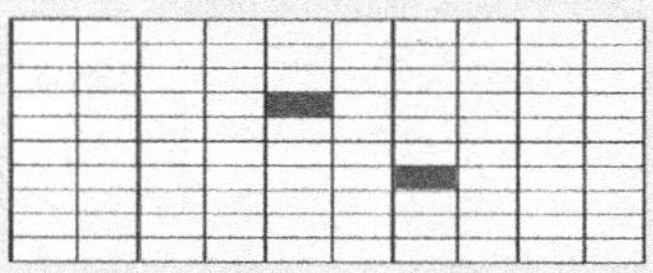	**2/100**	**Two hundredths**	**.02**	**2%**
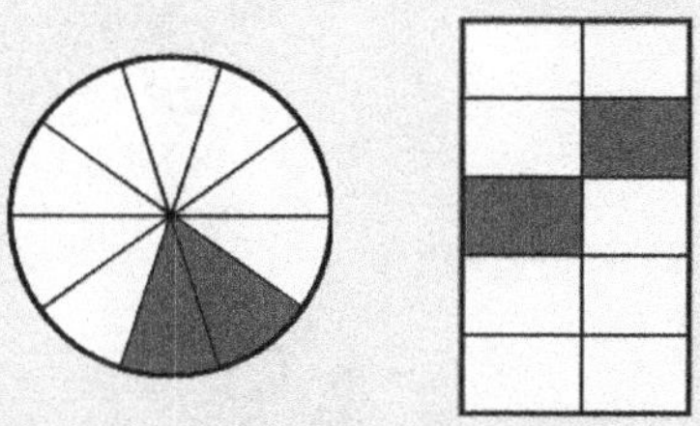	**2/10**	**Two tenths**	**.2 (or .20)**	**20%**
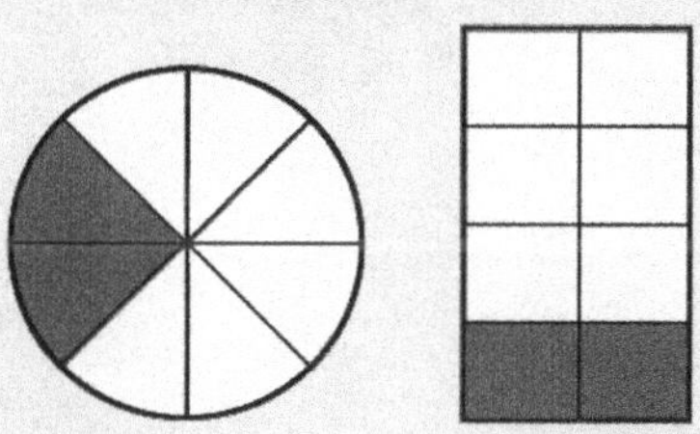	**2/8** *(same as ¼ or one fourth)*	**Two eighths**	**.25**	**25%**
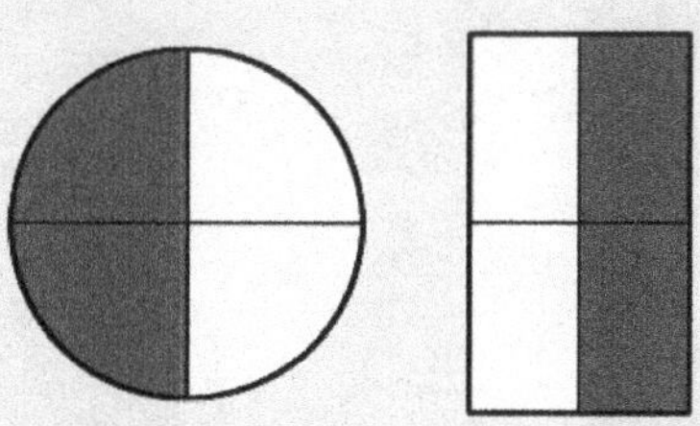	**2/4** *(same as ½ or one half)*	**Two fourths**	**.5 (or .50)**	**50%**
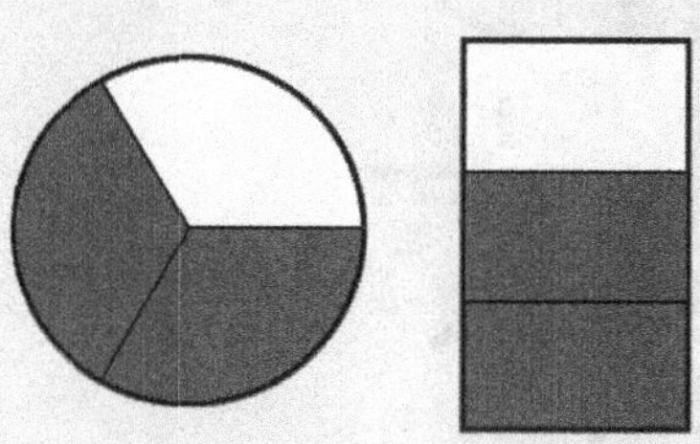	**2/3**	**Two thirds**	**.666666**	**66.66%** *(These repeat forever)*
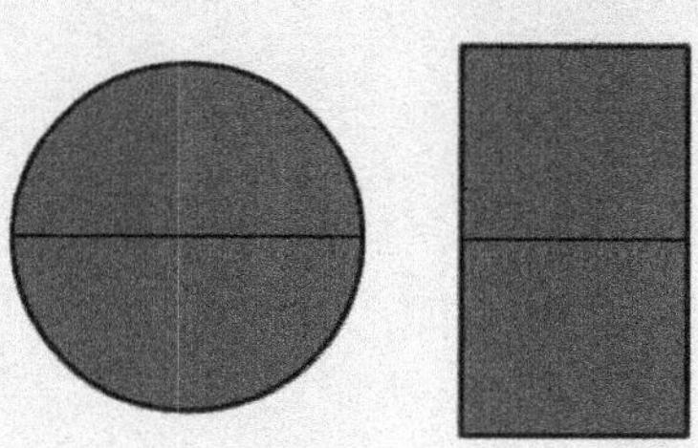	**2/2** *(This is also just called: 1)*	**Two halves**	**1**	**100%**

Decimals

Level 1

1) How do you say this number: **.2**
How do you write that as a fraction?

$$\frac{\quad}{10}$$

2) **1.1** candy bars is 1 whole bar plus ⅒ of a bar.

How many candy bars would you have if you had **1.2** candy bars?

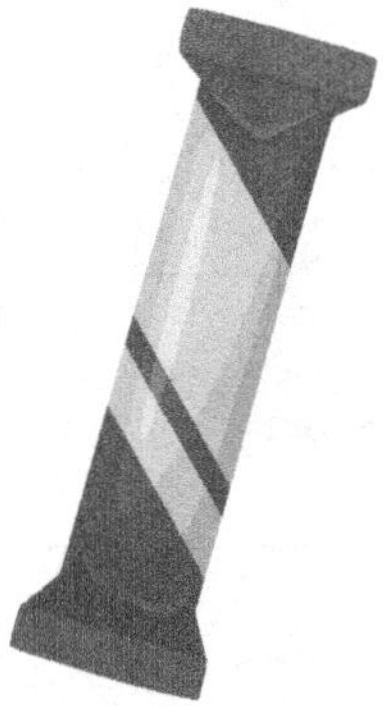

3) How do you say this number? **.07**
How do you write that as a fraction?

4) Shade in **.9** of this pizza. What decimal is not shaded?

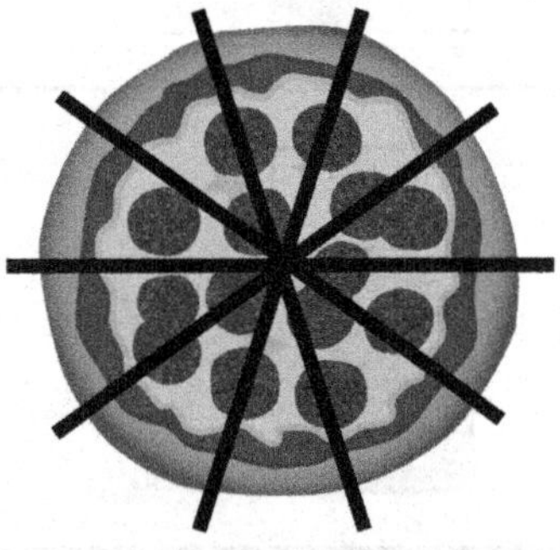

5) **½ 50% .5** are all different ways of saying "half" of something.
Circle the one that does NOT mean a "quarter" of something?

¼ 20% .25

6) Which decimal is larger? **.2** **.27**

7) Mark an **X** where you think the decimal **.5** would be:

8) What is **.2 + .5?**

9) Draw a picture of **3.5** frisbees:

10) What is **.05 + .35?**

Decimals

Level 2

1) How do you say this number: **.45**
   How do you write that as a fraction?

2) What are three different ways to describe this much pizza?

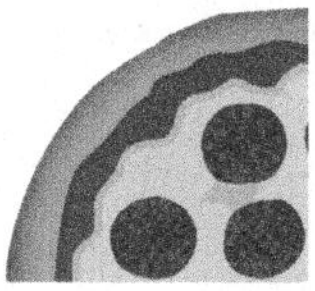

$\frac{\_\_\_}{4}$ ____% _._

3) Circle the larger decimal: **0.30** **0.29**

4) I ate **.4** of a pizza. How much is left?

5) Circle the numbers that mean a fifth: **⅕** **20%** **.5**

6) **.5 + __.___ = .80**

7) Mark an **X** where you think the decimal **.25** would be:

8) Put these in order from smallest to largest: **$\frac{3}{10}$** **.28** **.29**

9) Draw a picture of **3.75** plates.

10) What is **.56 + .44?**

Decimals

Level 3

1) Circle the larger decimal: **.2** **.18**

2) **.3 + .8 =** ____

3) Mark an **X** where you think the decimal **.333333333** would be:

4) How would you say this number: **0.375**
How do you write that as a fraction? ____

5) Order these from smallest to largest: $\frac{2}{10}$ **.03** **21%**

6) **2.9 + ____ = 3.1**

7) How would you show **$3.45** in expanded form (using money)?

8) **34.23 = 30 + 4 + $\frac{2}{10}$ + $\frac{3}{100}$**
Break apart this number in the same way: **56.011**

 + + ——— + ———

9) **.4 - .32 = ____**

10) Circle the number that is closest to 1:

$\frac{8}{9}$ 98% .099999 9.9% $\frac{9}{10}$ 0.9901

Decimals

Genius Level

1) Circle the smaller decimal: **0.6** **.09**

2) How would you estimate **3.334** pizzas as a fraction?

3) **.3 + .78 =** ____

4) Mark an **X** where you think the decimal **.999999** would be:

5) Mark an **X** where you think the decimal **0.0001** would be:

6) Circle the larger number: **.027** **.28%**

7) Put these in order from smallest to largest: **$\frac{3}{10}$** **.032** **33%**

8) Write this answer as a decimal: **3 + ½ + 20% + 0.2 =** ____

9) **$3.45** means 3 one-dollar bills + 45 pennies
If Chandra has **$13.03** and spends **$8.45,** how much money will she have left?

____ **dollars** ____ **cents**

10) Write **78.1234** in expanded form using fractions.

21

Estimation

Sometimes the exact answer in math is important, but other times all you need to do is get close. Look at this problem on the board.

I see what you mean. In this problem, Luke doesn't need to know the exact cost of the cantaloupes. He just needs to know how many he can buy with $20.

That would be easy to find. I'll just change the $1.95 to $2 because they are very close to each other and $2 is much easier to work with. Let's see, how many $2 fit into $20 —10!

Remember that sometimes an exact answer is important...

Tuna	*$2.13*
Ice Cream	*$4.92*
Bread	*$3.88*
Milk	*$4.92*

How much money should the clerk ask for?

A fancy word for making a good guess is "estimate." I could have easily estimated the cost of the 4 items and said the amount owed is $2 + $5 + $4 + $5 or $16. But stores need to know the exact amount to charge.

Estimating is really useful. I needed to know about how long a trip to my grandmother's house would take.

She lives 311 miles away and my family drives at a speed of 60 miles each hour. I changed the 311 to 300 and easily found that there are 5 groups of 60 in 300. Now I know my trip will take about 5 hours.

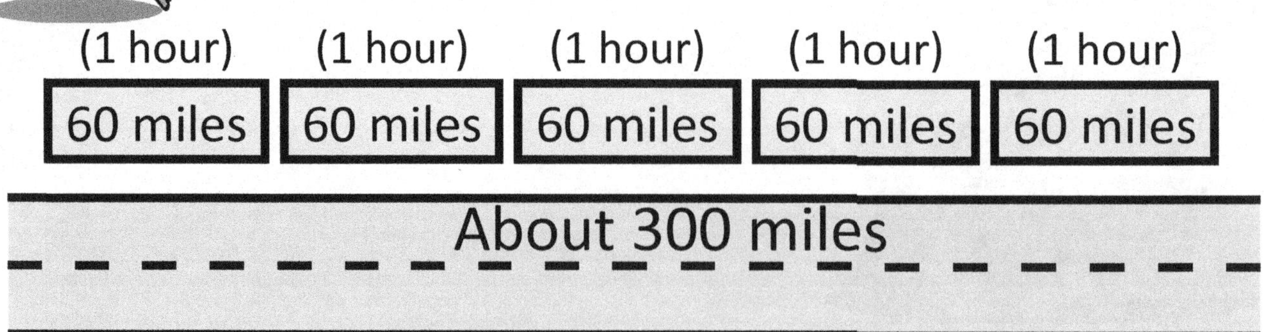

Estimating is also a great way to check if an answer makes sense. Pretend you have a problem on a test that is 501 x 12.

$$\begin{array}{r} 501 \\ \times \quad 12 \\ \hline 613 \end{array}$$

...you write down an answer. But does it make sense?

Estimate the answer by making the numbers easier: Change the 501 to 500 and the 12 to 10.

$$500 \times 10 = 5{,}000$$

The answer must be close to 5,000!

Your answer of 613 must not be correct, try solving it again!

Estimation

Level 1

1) Everet found $11 and wants to split it with his sister. **About** how much money should they each get?

2) Make a **good guess** for the cost of 6 ice cream cones that each cost $1.95.

$____

3) If $98 is to be divided among 4 people, **about** how much money will each person receive?

4) If you flip a coin 105 times, **about** how many times will the coin land on heads?

5) If you pay for $13.22 worth of groceries with a 20 dollar bill, **estimate** how much change you will get.

$____

Estimation

Level 2

1) What is a good **estimate** for the problem 103 x 11?

2) Debra has a box of 1000 apples. She took out 100 and saw 5 were rotten. **Estimate** how many rotten apples are in the full box of 1000.

3) Ed wanted to know about how much time his 8 mile walk would take. If the first mile took him 14 minutes, **estimate** how many minutes the entire walk would take.

4) **Estimate** the answer: $\frac{7}{8} + \frac{1}{2}$

5) If a car travels at a speed of 58 miles each hour, **estimate** how many hours a 180 mile trip will take.

Estimation

Level 3

1) **Estimate** the answer to the problem 1035 x 52.

2) If gas cost $2.48 each gallon, **about** how much does 8 gallons cost?

3) A bag of 1000 jellybeans contains 3 colors: Red, blue and green. If 100 jellybeans were picked and 98 were red, **estimate** how many of the total 1000 jellybeans are red.

4) If a car travels at a speed of 76 miles each hour, **estimate** how many hours a 600 mile trip will take.

5) **Estimate** how many hours it would take a plane to travel from Chicago to London (3951 miles) if it travels at a speed of 512 miles per hour.

Estimation

Genius Level

1) If you roll a die 592 times, **estimate** how many times the roll will be a "3".

2) The speed of sound is about 1100 feet per second. There are 5280 feet in one mile. **About** how long does it take for sound to travel one mile?

___ seconds

3) Jonathon wanted to know about how much time his 10 mile walk would take. If the first mile took him 14 minutes and 50 seconds, **about** how many hours will the 10 mile walk take?

4) **Estimate** how much faster light is than sound.

a) About the same speed
b) 1000 times faster
c) 100,000 times faster
d) 1,000,000 times faster

Light travels 186,000 miles per second.

Sound travels 1100 feet per second.

There are 5280 feet in one mile.

5) If the sun is 93,000,000 miles away, **about** how long does it take light to travel from the sun to the earth?

a) One minute
b) One hour
c) 8 minutes
d) 500 minutes

Part IV

Advanced Concepts

Knowing how to apply number sense opens the door to advanced and fun ideas...

22

Language of Algebra

Math has a very special language that I am going to teach you. It is called *algebra*.

My 12-year-old brother said he just started algebra in school. It sounds like it might be difficult, but I want to try!

When we have a mystery number, sometimes we put in a question mark like this problem on the blackboard. Because you are a little older, we are going to get fancier and put the letter *n* instead.

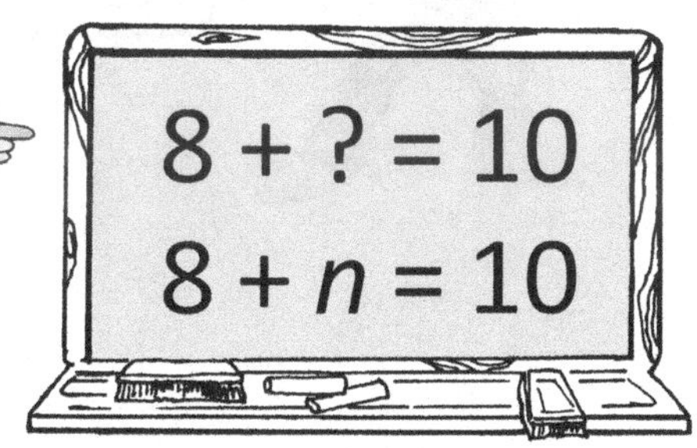

So, the *n* just takes the place of a number we don't know, just like a question mark. That doesn't seem too difficult. The answer here is pretty easy: 8 + 2 = 10. So the answer is *n* = 2

You solved that so easily that I think you are ready to learn the language of algebra. Once your brain gets used to the idea, I think you will find it fairly easy and also fun.

Watch how I turn this story about 2 brothers into the language of algebra.

Luke is 6 years older than Daniel.
I don't know Daniel's age.

I think I know how to do part of this. We do not know Daniel's age, so I will call Daniel's age *n*. That part was easy, but now I'm stumped... what do I call Luke's age?

If Daniel's age is *n*, then we have to look at what we know about the relationship between Luke's age and Daniel's age.

Luke is 6 years older than Daniel, so his age must be "*n* + 6"

Wow, if that is all there is to the language of algebra, I think I am ready for it.

If you understood how we changed ages into the language of algebra, then you are ready to jump right in. I want to warn you though, sometimes it gets a little more difficult and takes some clever thinking as we move forward. Good luck!

2 cows in a field have 2 x 4 = 8 cow legs.
If there are *n* cows in a field, how many cow legs are there?

This problem seems pretty easy. If there are 2 cows, all I have to do is multiply 2 x 4 = 8 cow legs because each cow has 4 legs. If the number of cows is *n*, then I need to multiply that mystery number by 4. The answer is 4 x *n*! We write "4 x *n*" like this: 4*n*

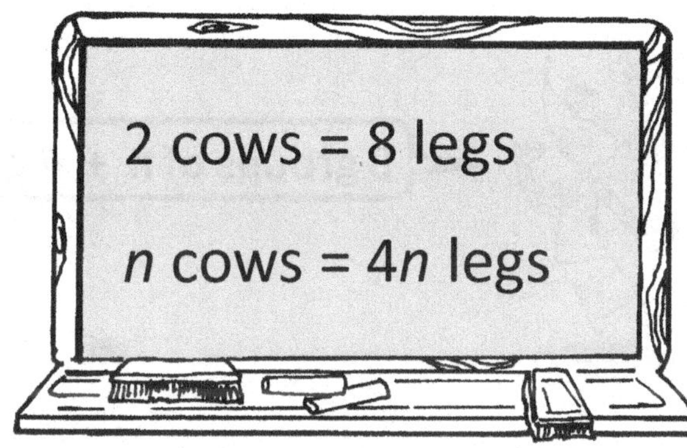

Emily is 5 years younger than her sister.
Emily is also 10 years younger than her brother.

If we call Emily's age *n*, what do we call the ages of her sister and brother in the language of algebra?

This problem is a little harder. Emily is 5 years younger than her sister, which means her sister is 5 years *older* than Emily. I will call her sister's age *n* + 5.

Now it is pretty easy to see that her brother's age is *n* + 10 because he is 10 years older than Emily.

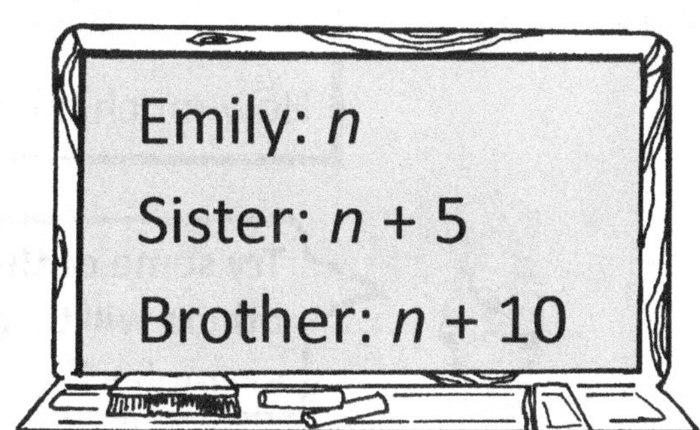

There are the same number of spiders and mice in a box.
If we call the number of spiders *n*, how many total legs are in the box?

Hmm because the number of spiders is *n*, there must be 8 x *n* spider legs, so 8*n*. (Putting the number next to the letter means the same as multiplying.)

I know there are also *n* mice in the box, so the number of mouse legs must be 4 x *n* or 4*n*.

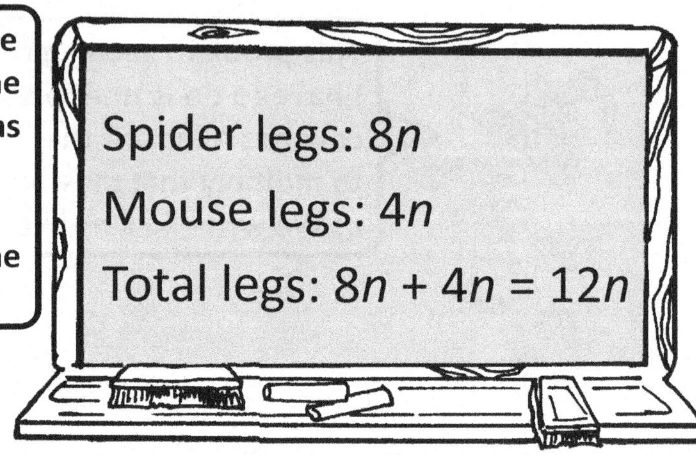

8 groups of *n* + 4 groups of *n* = 12 groups of *n*

Here are a few more examples...

How many spider legs are there in a box with *n* spiders? **8*n* legs**

How many duck legs are there in a box with *n* ducks? **2*n* legs**

The total cost of *n* candy bars that are \$2 each: **\$2*n***

How many inches are there in *n* feet? **12*n* inches**

A spider collection has *n* legs, how many spiders are there? **$n \div 8$**

How much money is in a box with *n* quarters? **25*n* cents**

Try some of the problems on the next few pages. Some will be easy and others will take a lot of thinking. Look at the chart above to see different examples if you get stuck.

Language of Algebra

Level 1

1) David is *n* years old. Alexis is 3 years older than David. What is Alexis' age in the language of algebra?

2) A tree in Emma's yard is 20 feet taller than Emma.
If Emma is *n* feet tall, how tall is the tree?

3) Daphne found a box of money that had *n* dollars in it. After she spent $10, how much money is left in the box?

4) Jamie counted the rings on a slice of an old fallen tree to find out how old it was. He said the tree is "15 years older than my grandfather." If we call his grandfather's age *n*, how old is the tree in the language of algebra?

5) 5 ducks x 2 legs = 10 duck legs. 2 cows x 4 legs = 8 cow legs.
If there are *n* ducks swimming in a lake, how many duck legs are there on the lake?

Language of Algebra

Level 2

1) If a fox walked for 3 minutes, that means it walked for 180 seconds (3 minutes x 60 seconds per minute = 180 seconds).

   If the fox walks for *n* minutes, how many seconds did it walk?

2) A box of 6 ladybugs has 36 legs (6 ladybugs x 6 legs each = 36 legs). How many legs does a box of *n* ladybugs contain?

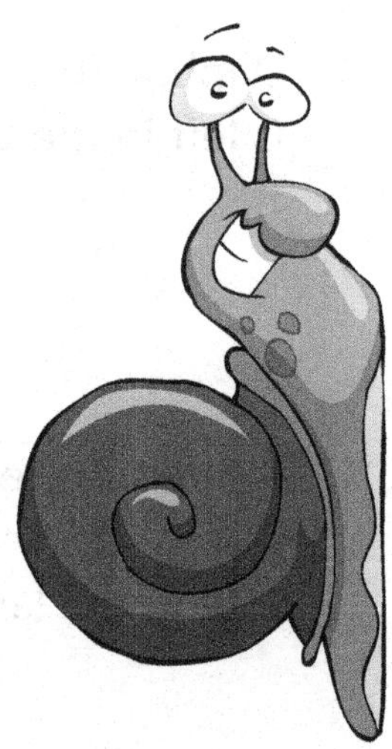

3) If snails cost $2 each, what is the cost of *n* snails?

4) A gallon contains 8 pints. 4 gallons is equal to 4 x 8 = 32 pints. How many pints are in *n* gallons?

5) The value of 5 dimes is 10 cents x 5 = 50 cents. Mike found a box of dimes but doesn't know how many are in the box. The number of dimes in the box is *n*. What is the value of the dimes in the box?

Language of Algebra

Level 3

1) Henry weighs 5 pounds more than twice his dog's weight. If Henry's dog weighs *n* pounds, how much does Henry weigh in the language of algebra?

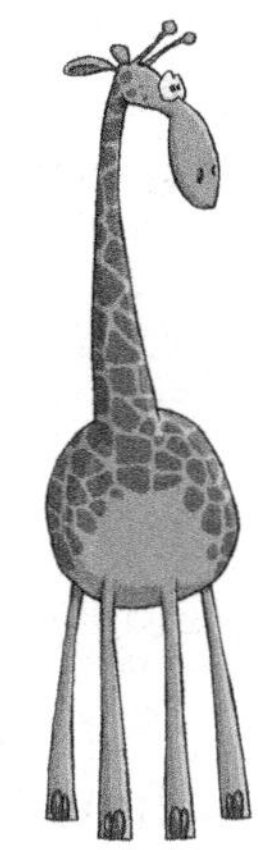

2) A giraffe collection at a zoo has *n* legs. How many giraffes are in the collection?

3) If a rat ran for *n* yards, how many inches did the rat run? (*There are 36 inches in a yard.)*

4) A barn has several animals in it. It has *n* pigs, 3 ducks, and 2 people. How many legs are in the barn?

5) What is the value of a box that holds *n* quarters and *n* nickels?

Language of Algebra

Genius Level

1) A group of *n* horses, *z* cows, and *y* ducks went to a movie theater. The owner said they would need to pay $3 per leg. How much money did their tickets cost?

2) If a bus drove for *n* minutes, how many hours did it drive?

3) A field has a total of 100 animals in it (only cows and horses). If there are *n* horses, how many cows are there in the language of algebra?

4) Samantha is 3 times older than her pet rat. The rat is twice as old as Samantha's brother. If Samantha's brother is *n* years old, how old is the rat? How old is Samantha?

Rat's age:
Samantha's age:

5) There are twice as many nickels as quarters in a box. What is the total value if there are *n* quarters in the box?

23

Balancing Equations

Solving equations is all about balance!

I want both sides of this scale to stay balanced. Remember that *n* stands for a number we do not know.

$$n + 8 = 12$$

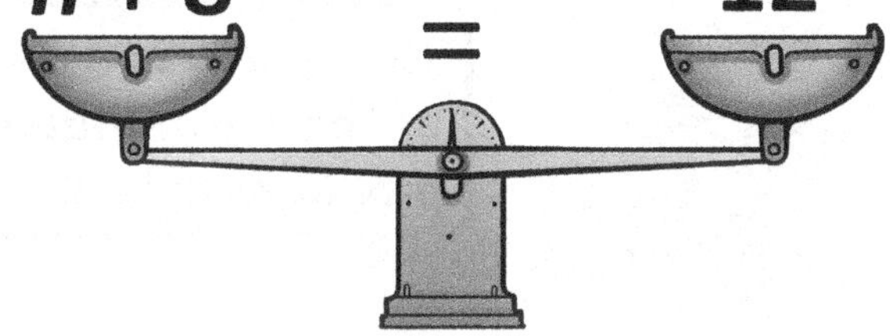

That seems easy.
"Some number" plus 8 is 12.
The mystery number must be 4!

This one is a little more difficult, but you can still solve it in your head. "Some number" minus 8 is 2.

$$n - 8 = 2$$

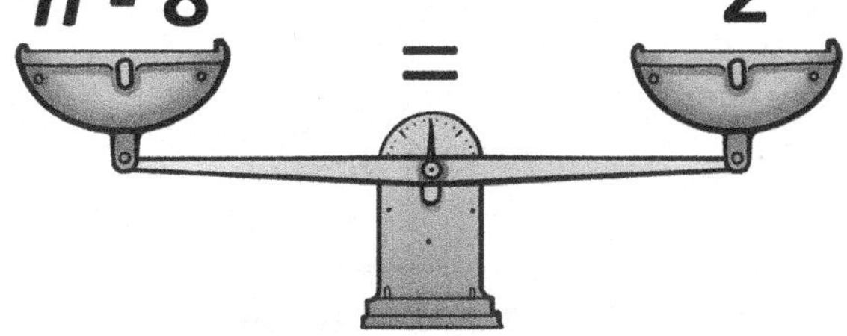

You're right, that is more difficult, but there is only one answer that fits. The mystery number must be 10!

This one is a little different. I know that 3*n* is just a shorter way of saying "3 x *n*" or "3 groups of *n*".

Let me take a few guesses and see if I can find what number times 3 is equal to 21.
3 x 5 = 15
3 x 6 = 18
3 x 7 = 21

That wasn't too hard, the answer is 7!!

The "guess and check" strategy works for some of the easier equations, but what about when problems get a little more complicated?

Multi-Step Balancing

This equation looks like one of the more complicated ones!

$2n + 4 = 14$

I tried to guess and check in my head, but I kept getting confused.

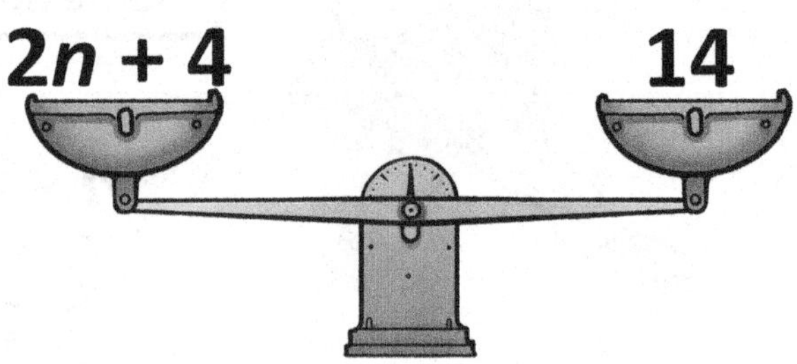

Let's pretend the numbers are about bricks to make them easier to think about.

We will call *n* a mystery bag of bricks, so "2*n*" means "2 mystery bags of bricks." Look at the picture on the next page. We are trying to find out how many bricks are in each bag (*n*).

2n + 4 bricks = 14 Bricks
n
n
=
I want to find out how many bricks are in each bag (n) so I need to get rid of these 4 bricks.
n
n
=
Oh no, now the scale isn't balanced! I just took off 4 bricks from the LEFT side, so maybe I should also take off 4 bricks from the RIGHT side.
n
n
=
All balanced. I feel much better now!
Now I see that the 2 mystery bags of bricks equals 10 bricks:
2n bricks = 10 bricks
That means I can take half of both sides to find my answer!
n
n
=
n = 5 bricks

Let's do the same thing again with only numbers this time. Remember we need to get the *n* all by itself on one side of the equation. Sometimes this takes a few steps.

$$2n + 4 = 14$$

(subtract 4 from both sides)

$$2n + 4 \overset{-4}{} = 14 \overset{-4}{}$$

(2 groups of n equal 10...)

$$2n = 10$$

I want to know how much is ONE group of n

$$2n \overset{\div 2}{} = 10 \overset{\div 2}{}$$

n equals 5

$$n = 5$$

***n* is by itself and the scale is still balanced. That means there are 5 bricks in each group.**

I can plug the answer into the original equation to make sure it is correct!

5

$$2n + 4 = 14$$

$$10 + 4 = 14$$ ✓

All by *n*'s Self

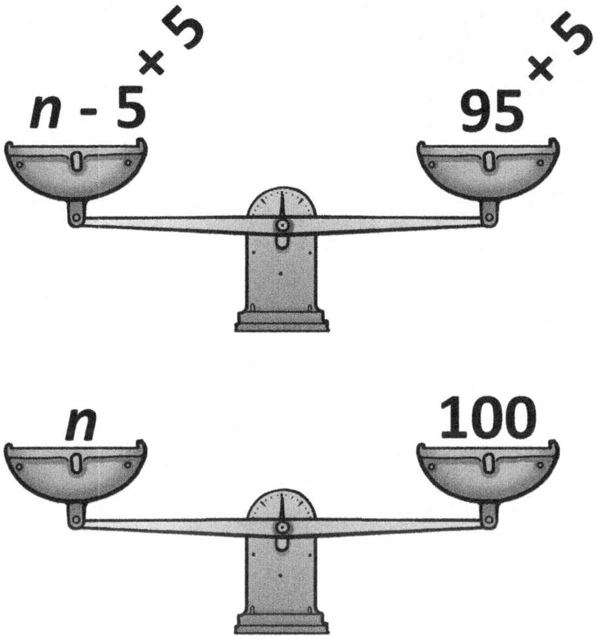

Always try to get the *n* all alone on one side of the equation. If a problem is $n - 5 = 95$ we get rid of the "-5" by adding 5.

That's because combining -5 and +5 gives you 0!

Watch how I solve the problems below, then you will be ready to try some on your own! It is great if you can do them in your head, but as they get harder you might need to write down each step as you try to get *n* by itself.

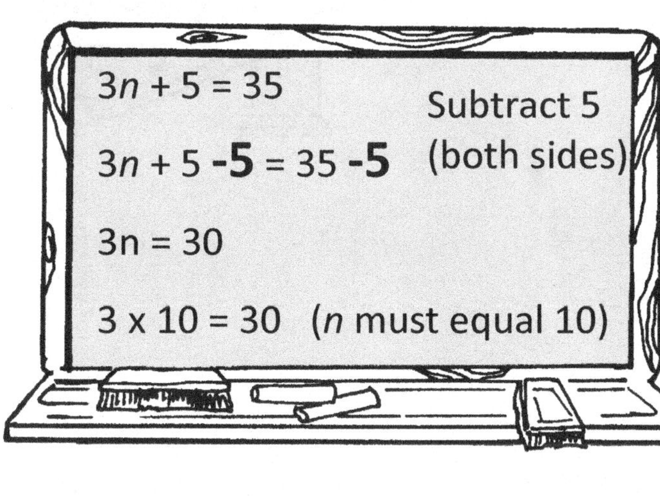

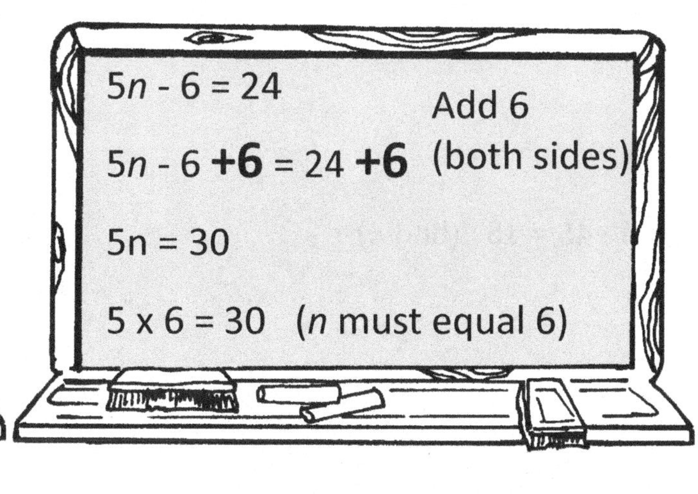

Balancing Equations

Level 1

1) What number is n?

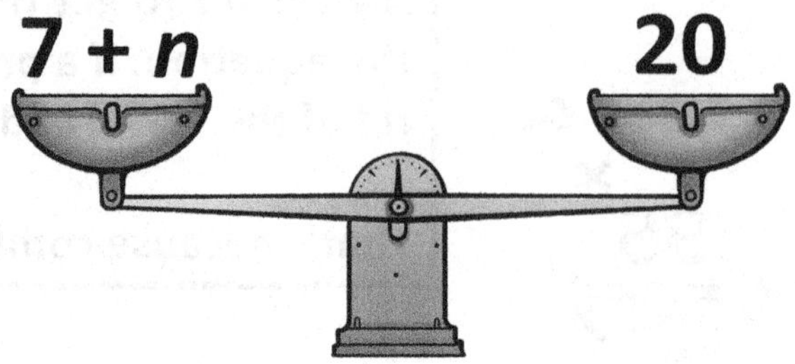

2) $18 + n = 30$ (find n)

3) $n - 90 = 10$ (find n)

4) How many bricks should be put on the left side to balance the scale?

5) $n - 45 = 15$ (find n)

6) $12n = 0$ (find n)

7) $n + 500 = 500$ (find n)

8) $10n = 10$ (find n)

9) $10 - n = 4$ (find n)

10) $12 - n = 12$ (find n)

Balancing Equations

Level 2

Find n for the following problems...

1) $5n = 45$

2) $10n + 1 = 51$

3) $100 - n = 10$

4) $10n + 100 = 100$

5) $12n + 3 = 15$

6) $10n - 1 = 49$

7) $8n - 10 = 6$

8) $4n - 5 = 95$

9) $4n + 82 = 182$

10) $12n + 1 = 145$

Balancing Equations

Level 3

Find n for the following problems...

1) $1n = \$1$

6) $12n = 6$

2) $n - 11 = 20$

7) $25n + 17$ cents $= 67$ cents

3) $6n + 100 = 103$

8) $\frac{1}{2}n + 50 = 550$

4) $2n + 10 + 3n = 35$

9) $100 - 2n = 10$

5) $9n + \$10 = \$14.50$

10) $n + n + 2n = 1$

Balancing Equations

Genius Level

1) $99 + 4n = 100$ (find n)

2) $1 - n = \frac{1}{8}$ (find n)

3) $24 - 3n = 6$ (find n)

4) A box of spiders had 96 legs.
What equation should be written if you call the number of spiders n?
How many spiders are in the box?

5) Jack and Jill are the same age. If you add their ages and then added 8 more, you would get 40.
What equation should you write? How old is Jack?

Jack's Age: n
Jill's Age: n

6) There are a total of 212 animals in a field. These include 20 blue goats, 20 spotted goats, and 2 equal groups of cows. How many cows are in each group?

7) Find n: $10n + 985 = 990$

8) Find n: $2n - 18 + 3n = 42$

9) Steven is 4 years older than Tyler. If their ages add up to 36, how old is Tyler?

10) A pile of n quarters and n nickels is worth the same as a pile of 99 dimes. How many quarters are there?

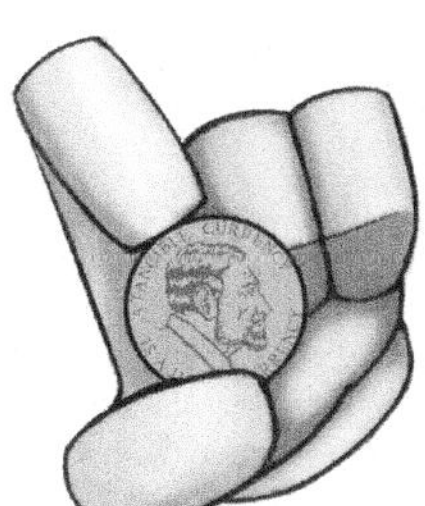

24

Ratios

Ratios are all about how two groups relate to each other. Think of them like this:

how many ____ per ____

That reminds me of my grocery store! They sell boxes of fruit with 2 bananas and 4 apples. So for every group of 2 bananas, we will also have a group of 4 apples.

That's right, so we can say we have a ratio of 2 bananas to 4 apples. In math we show ratios like this...

2 bananas : 4 apples

RATIO
2 bananas : 4 apples

2 bananas, 4 apples

4 bananas, 8 apples

6 bananas, 12 apples

Moving Backwards

RATIO
2 bananas : 4 apples

1 box  **2 bananas, 4 apples**

½ box  **1 bananas, 2 apples**

¼ box  **½ banana, 1 apple**

Ratios can also move the opposite way. If we had a HALF box of fruit, the relationship between bananas and apples would stay the same!

This reminds me of the double number line from our Percents chapter. The relationship between bananas and apples stays the same, so I can move this slider left and right to show how many of each fruit I would get!

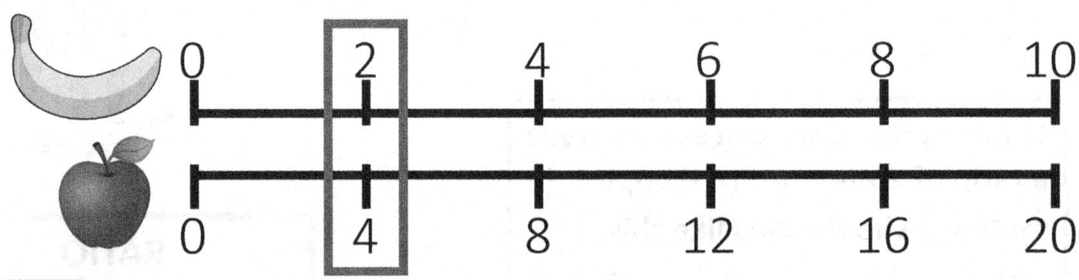

Notice how the zeroes line up. Zero bananas also means zero apples!

BANANAS	APPLES
1	2
2	4
3	6
4	8
5	10

You could also make a table. Tables are more abstract because they don't represent size or length like a number line. But this also makes them faster to write!

Real-Life Ratios

Ratios happen in real life all the time...

...and they can even include more than two things!

Cats

2 eyes : 4 legs

Spiders

8 eyes : 8 legs

Recipes

1 cup flour: ½ cup sugar: 3 eggs

Some ratios are easy to see when the sun comes out!

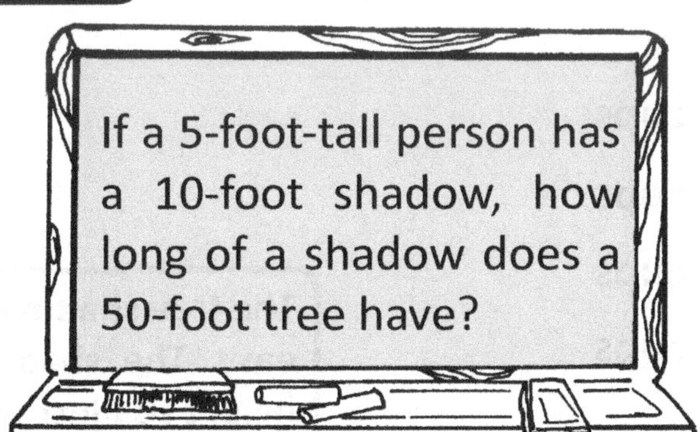

5 foot (person) : 10 foot shadow

50 foot (tree) : ___ foot shadow?

I don't even need my calculator for that problem. The person's shadow is twice as tall as the person, so the tree's shadow must be twice as long as the tree.

The tree's shadow must be 100 feet long!

Ratios are especially important for recipes. Look at these instructions for making lemonade...

1 scoop of lemonade mix FOR EVERY 2 cups water

My pitcher holds 10 cups of water. How many scoops of mix do I need for a full pitcher?

I know it is important to keep the relationship between "scoops" and "water" the same so my lemonade ends up tasting right!

1 scoop mix : 2 cups water

___ scoop mix : 10 cups water?

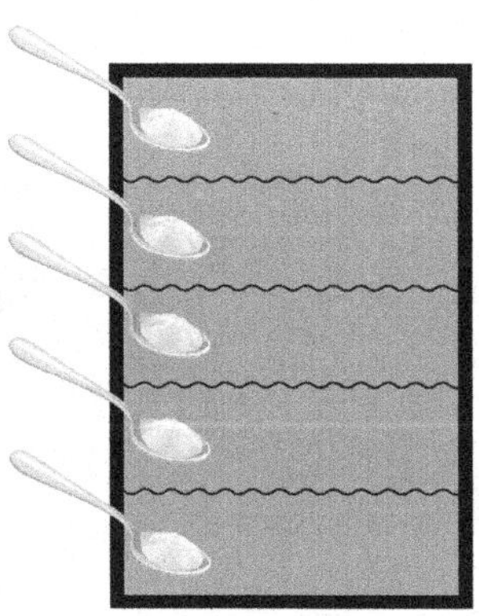

The tree shadow problem was super easy. The shadow was double the height so I just multiplied by 2...

...but the lemonade problem is a little harder. I need to figure out how many groups of 2 fit into 10 cups of water...

2 2 2 2 2

5 groups of 2 (5x2) is a total of 10 cups!

That means I need to follow the recipe 5 times, so I need 5 scoops of lemonade mix. Turns out I really like ratios!

Unit Price

Finding the "unit price" or "how many per ONE" is a great way to help solve ratio problems. If you are having trouble finding an answer, try dividing the ratio until you have "one" of something.

Ratios

Level 1

1) Gabe is 4 feet tall and has a shadow of 8 feet. If his mom is 5 feet tall, how long will her shadow be?

2) Jeremy is 4 feet tall and has a shadow of 8 feet. If a tree in his yard has a shadow of 100 feet, how tall is the tree?

3) A 10-foot post had a shadow of one foot. How long of a shadow would a 100-foot tall building have?

4) Natalie's allowance is double her brother's allowance. If her brother's allowance is $6.50, what is Natalie's allowance?

5) Natalie's allowance is still double her brother's allowance. If Natalie's allowance increases to $22, what is her brother's allowance?

6) A recipe for 100 cookies calls for 4 cups of flour. If you want to make only 50 cookies, how much flour will you need?

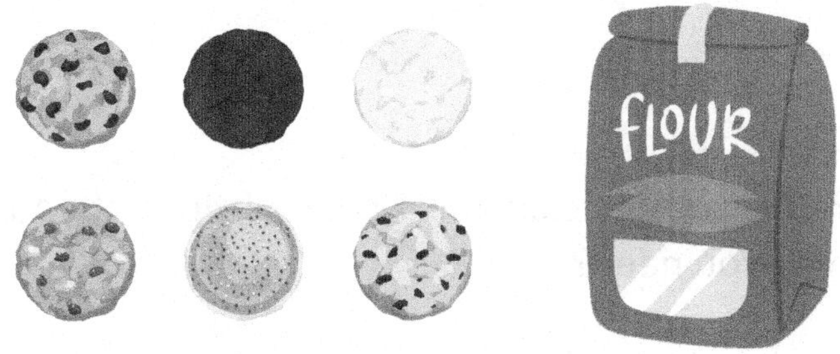

7) A 273-foot-tall cell tower has a shadow of 273 feet. How long will Bill's shadow be if he is 4 feet and 3 inches tall?

8) If a bag of dog food will feed 4 dogs for 2 weeks, how long will the same bag feed 8 dogs?

9) A map has a scale of 1 inch = 150 miles. If two cities are 2 inches apart on the map, how many miles apart are they?

10) If a one-inch-tall grasshopper can jump 10 inches, how far could a 3-foot-tall child jump if she had the ability to jump like a grasshopper?

Ratios

Level 2

1) A recipe for 100 cookies calls for 4 cups of flour. If you want to make only 25 cookies, how much flour will you need?

2) A 5-foot post has a shadow of 3 feet. How long of a shadow does a 10-foot tree have?

3) Jenna's allowance is triple her brother's allowance. If her allowance is $24, what is her brother's allowance?

4) Eric has a box of 1000 marbles that are either blue or red. He closed his eyes, picked out 100 marbles at random, and saw there were 50 blue marbles and 50 red marbles in his hand. About how many of Eric's 1000 marbles do you think are blue?

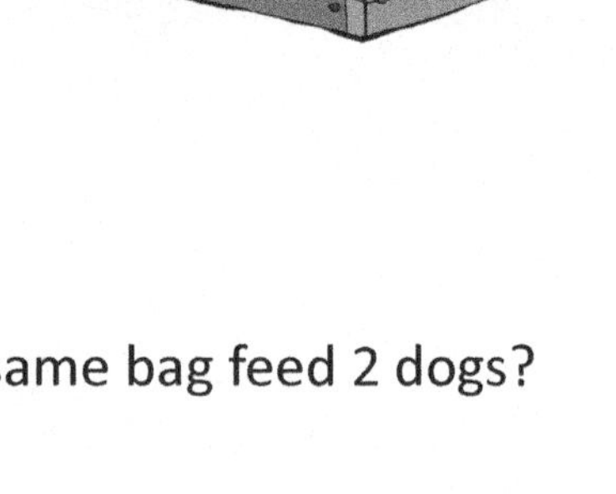

5) If a bag of dog food will feed 4 dogs for 2 weeks, how long will the same bag feed 2 dogs?

6) A map has a scale of 1 inch = 25 miles. If two cities are 4 inches apart on the map, how many miles apart are they?

7) A bug that weighs one ounce can carry 5 ounces of food. If a child who weighed 50 pounds had the ability to carry weight like the bug, how much weight could he carry?

8) If an upright yardstick has a shadow of 18 inches, how tall is a tree with a shadow of 10 feet?

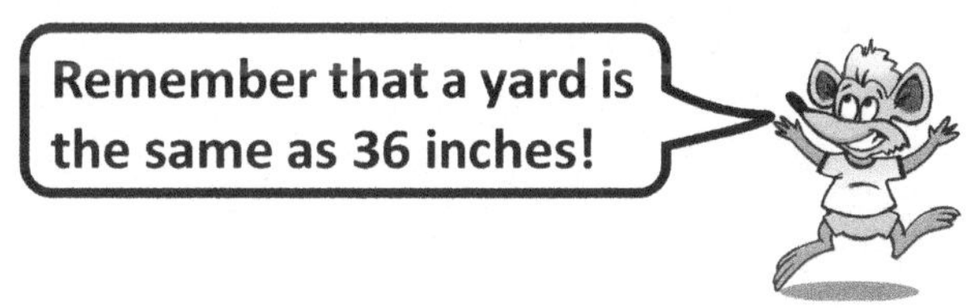

9) A cookie recipe that makes 100 cookies calls for 8 eggs and 2 pounds of flour. If Ethan wants to make 150 cookies, how many eggs and how much flour should he use?

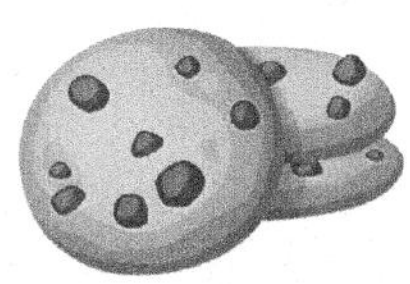

10) Jayden had a 60-pound statue that was made of 20 pounds of gold and 40 pounds of silver. Jayden broke off the statue's leg and found it had one pound of gold in it. How many pounds of silver is in the statue's leg?

Ratios

Level 3

1) If an upright 12 inch ruler has a shadow of 4 inches, how long will an upright yardstick's shadow be?

2) A recipe for 100 cookies calls for 4 cups of sugar. If you want to make 250 cookies, how much sugar will you need?

3) Natalie's allowance is 4 times her brother's allowance. If Natalie's allowance is $10 each week, what is her brother's allowance?

4) If an upright 10-foot tall stick has a shadow of 15 feet, how tall is a tree that has a shadow of 75 feet?

5) The grocery store sells large boxes of fruit, each with 20 bananas, 5 apples, and some kiwi's. You buy 3 boxes and count a total of 84 pieces of fruit. How many kiwi's were in each box?

6) Jayden had a 100-pound statue that was made of 10 pounds of gold and 90 pounds of silver. Jayden broke off the statue's leg and found out it weighed 10 pounds. How many pounds of silver are in the statue's leg?

7) 1000 students voted on a new name for their school. The choices were "Owl Elementary" or "Eagle Elementary." Bradley wanted to predict what name would win, but he didn't want to call 1000 children to ask them how they voted.

What do you think the total vote will be for all 1000 students?

Owl Votes:

Eagle Votes:

8) If 1 bag of dog food will feed 5 dogs for 5 weeks, how long will the same bag feed 1 dog?

9) A 10-foot flagpole has a shadow of 2 feet. How tall is a tree with a shadow of 10 feet?

10) A map has a scale of ½ inch = 100 miles. If two cities are 500 miles apart, how far apart are they on the map?

Ratios

Genius Level

1) Lebron launched a rocket straight up and measured the shadow of the rocket as it went along the ground. He found that the rocket's shadow traveled 100 feet. If Lebron is 4 feet tall and had a shadow of one foot, how tall did the rocket go?

2) A recipe for a large 12-pound cake calls for one gallon of milk. If you want to make a 3-pound cake, how many quarts of milk should you use? *Hint: There are 4 quarts in a gallon.*

3) Kim's 30-foot tall house has a shadow of 5 feet. If Kim's shadow is 10 inches, how many feet tall is Kim?

4) Complete the missing values in this table:

Mix	Water
0	___
½	¼
3	___
⅘	___
___	1

5) Explain why ratios are harder to understand with a **table** compared to a **double number line**.

Table

Double Number Line

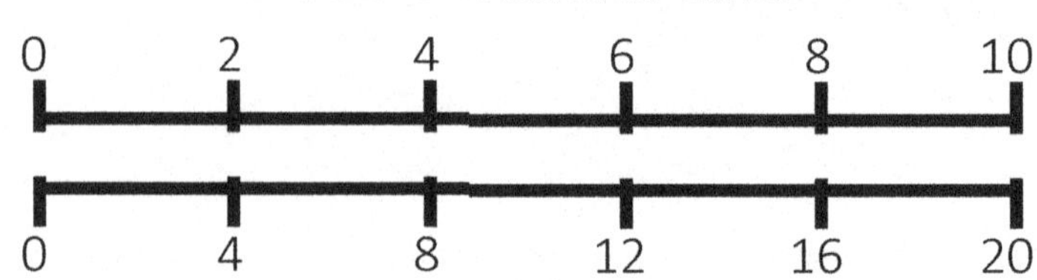

6) Fill in the numbers on the top of this Double Number Line:

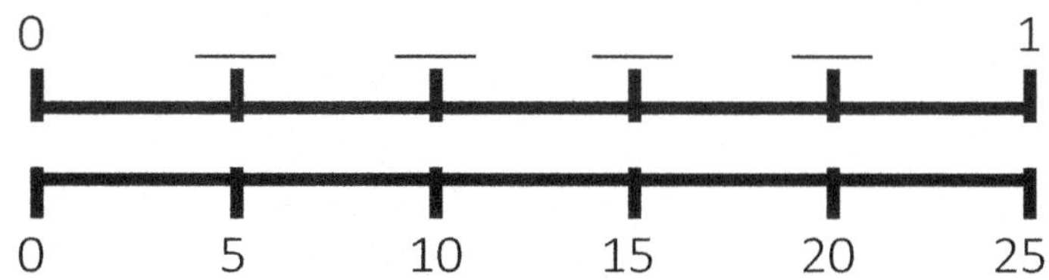

7) 8 people have enough food for a 6 day hike. If 8 more people join them, how many days will the food last?

8) A large globe has a scale of 2 inches = 1000 miles. If two cities are 7 inches apart on the globe, how many miles apart are they?

9) The tallest redwood tree grows in California and is about 360 feet tall. How long of a shadow would this tree have at the time a 4 foot tall person casts a shadow of 5 feet?

10) Jack climbed to the top of the beanstalk and saw an enormous giant. He wanted to know how tall the giant was, so he quietly measured the giant's shadow and found that it was 20 feet long. Jack measured himself as 5 feet tall, then measured his shadow as 2 feet long. How tall was the giant?

25

Speed, Distance, and Time

Speed is a special kind of ratio, and it usually looks like "miles PER hour" (MPH). Look at the speeds of these animals. The fastest animal is the cheetah, which can run at a speed of 75 mph.

Cheetah	75 mph
Antelope	60 mph
Lion	50 mph
Hare	50 mph
Sloth	$\frac{1}{10}$ mph
Kangaroo	44 mph
Snail	$\frac{1}{2}$" per second

Cheetahs are fast, but my favorite animal is the lion. They can run 50 mph, but I'm not sure what that means...

Remember how we used ratios to say this?

2 Bananas : 4 Apples

Speed works the same way, so we can think of 50 miles per hour like this...

50 miles : 1 Hour

So for every hour the lion runs, it can travel 50 miles. For example, that's 50 miles for 1 hour, 100 miles for 2 hours, 150 miles for 3 hours, and so on...

That makes sense! That also means if a lion ran at 50 mph for *half* of an hour, it would travel *half* of 50 miles, so 25 miles. Wow, that means it will take the snail a long time to go one mile!

Next time you are riding in a car, look at the speedometer. It tells you how fast the car is traveling. Knowing your speed allows you to predict how long your trip will take.

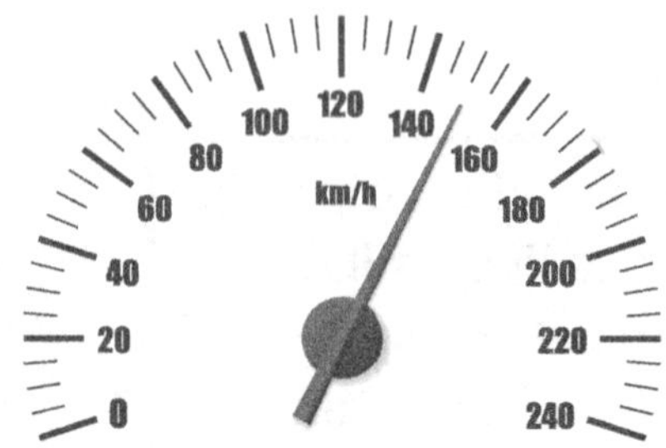

If your Grandma lives 300 miles away, how long will it take to get there if you drive 60 miles per hour?

That seems pretty easy. If we drive 60 miles each hour, it will take me 5 hours to travel the 300 miles. Now I can tell my brother how long the trip will take so he stops asking "are we there yet?"

60 miles	60 miles	60 miles	60 miles	60 miles
1 hour	**1 hour**	**1 hour**	**1 hour**	**1 hour**

Number Line Road

Double number lines are even more helpful when we're looking at distance! Think about them like the road we are traveling on.

Look out for cars, Mr. Lion!!!

At a speed of 50 miles per hour (MPH), how far will a Lion run in 1½ hours?

That's an easy way to visualize how fast a lion can run each hour!

I like to draw rectangles to solve distance & time problems. It lets me chunk pieces of time together so I can understand the distance better.

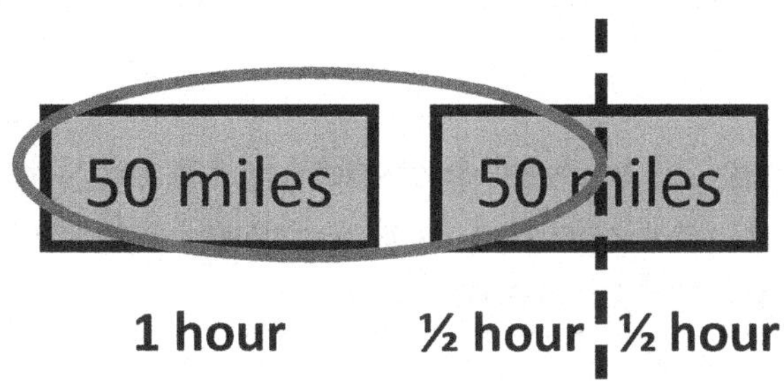

Finding the Speed

Some problems are reversed and give us the distance instead of the speed.

Hmmm, I know there are 12 inches in a foot. If the snail is moving in chunks of 1 inch, that means it will move 1 inch a total of 12 times. If each 1-inch chunk takes 2 seconds, that means it will take the snail 24 seconds to travel a foot!

I'll draw a few squares to check my answer.

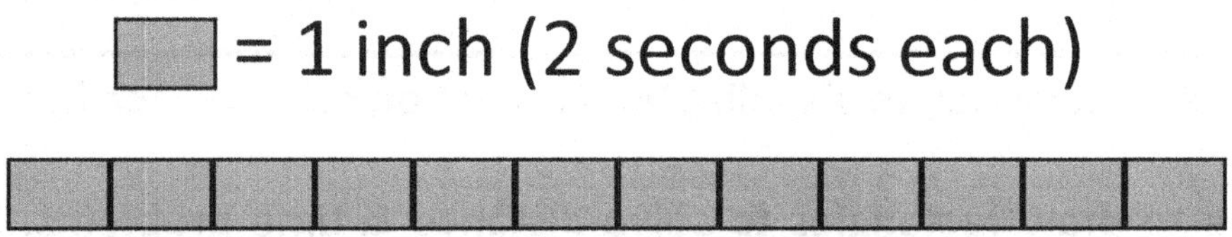

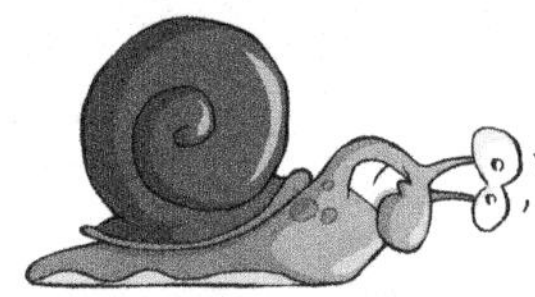

Instead of drawing shapes, you can set up an equation with the numbers and find the "unknown." We can write ratios as fractions when we use them in equations like this.

Notice how we put "12 inches" instead of "1 foot." This is because the top row and the bottom row each need to use the same kind of measurement.

$$\frac{1''\ \text{(Inches)}}{2\ \text{(Seconds)}} = \frac{12''\ \text{(Inches)}}{n\ \text{(Seconds)}}$$

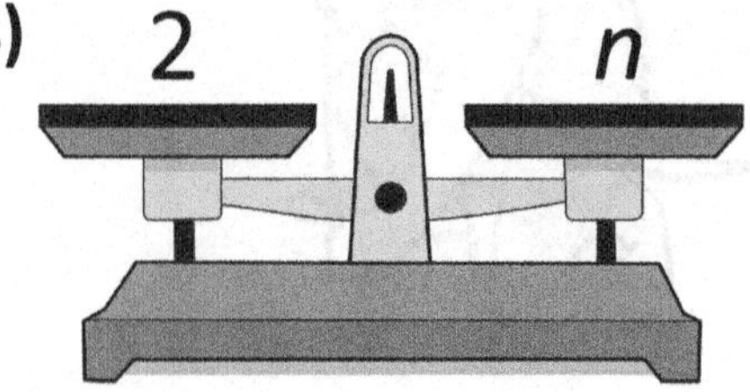

Now that we have an equation, all we need to do is find *n*!

We could try to get *n* by itself on one side, or we could think about the ratios to find the answer.

Looking at the TOP part of the equation, the 12 is "12 times bigger" than the 1.

The BOTTOM part needs to balance, so that means the *n* will have to be "12 times bigger" than the 2.

(x12)

$$\text{(Inches)}\ \frac{1''}{2}\ \text{(Seconds)} = \frac{12''}{n}\ \text{(Inches)}\ \text{(Seconds)}$$

(x12)

These kinds of equations can be solved even faster with "Cross Multiplying." That means we multiply the diagonal numbers we know, then divide that answer by the third number we know.

$$\frac{1}{2} = \frac{12}{n}$$

1. Multiply the diagonal numbers we know — **12 x 2 = 24**
2. Divide by the third number — **12 ÷ 1 = 24**

Speed, Distance, and Time

Level 1

1) If a wolf walked 10 miles in 2 hours, how many miles did it walk each hour?

2) A family went on a 10-mile hike and it took them 5 hours. What was their speed in miles per hour?

3) Lions can run for short distances at a speed of 50 miles per hour. If a lion was able to run at that speed for half of an hour, how far did it run?

4) A marlin can swim at speeds close to 25 miles per hour. If a marlin swims at that speed, how far can it travel in 4 hours?

5) The Blue Wildebeest can run at a speed of about 50 miles per hour, which matches the top speed of a lion. If a car travels at a speed of 50 miles each hour, how far will it go in 10 hours?

Speed, Distance, and Time

Level 2

1) If a hare was able to jump at 48 miles per hour for half an hour, how far did it jump?

2) The Galapagos tortoise is the largest tortoise species and can live for 150 years. Their top speed is about 1 mile per hour. If a Galapagos tortoise moves at that speed, how much time will it need to travel 3½ miles?

3) If a whale swam 120 miles in 10 hours, how many miles per hour did it travel?

4) A family decided to drive 660 miles for their vacation. If they travel at 60 miles per hour, how long will it take them?

5) The fastest sprinters in the world can run at a speed of about 25 miles per hour, but they can only run at that speed for 10 to 20 seconds. If they were able to run at that speed for 500 miles, how long would it take them?

Speed, Distance, and Time

Level 3

1) Cheetahs can run for short distances at a speed of 75 mph. How far can they run in 20 minutes?

2) The common garden snail has a foot-like organ that moves the snail very slowly. To help it move, the snail releases mucus to help reduce friction (that is why snails leave a trail of slime). A garden snail's top speed is only ½ inch per second. At that speed, how long will it take a garden snail to travel three feet?

3) If a toddler walked ¼ of a mile in 15 minutes, what was his speed in miles per hour?

4) The peregrine falcon is not only the fastest flying bird, but also the fastest animal on the earth. When it starts its hunting dive, it soars to a great height and then dives at a speed of 200 miles per hour. At this speed, how far can it travel in 15 minutes?

5) Light travels at a speed of about 186,000 miles per second. The moon is about 250,000 miles away. How long does it take for light to travel from the earth to the moon?

a) Less than a second
b) About one minute
c) A little more than one second
d) A little more than one hour

Speed, Distance, and Time

Genius Level

1) When a peregrine falcon dives, it can travel at a speed of about 200 miles per hour. How far can a falcon travel in one minute?

a) About one mile b) About 3 miles c) About 10 miles

2) The international space station orbits the earth 16 times each day as it travels at a speed of 5 miles each second. How many miles per hour is this? *(Remember there are 60 minutes in each hour and 60 seconds in each minute.)*

3) Sloths live in the treetops and move very slowly because they have a low metabolic rate. They move so slowly that sometimes algae grows on their fur. If a sloth moves at a speed of 8 feet per minute, how far will it move in one hour?

4) Light travels at a speed of about 186,000 miles per second. The sun is about 93,000,000 miles away. How long does it take for light to travel from the sun to the earth?

a) About one second
b) About 8 minutes
c) 93,000,000 seconds
d) A little more than one hour

5) A garden snail's top speed is only ½ inch per second. At that speed, how many minutes will it take a garden snail to travel from home plate to 1st base in a baseball game (90 feet)?

26

Negative Numbers

So far we've been working with numbers greater than or equal to zero. But look at these strange numbers...

-10

-15

-20

I know the "minus symbol" means subtract, but subtract what?

I've seen these numbers before! They are called negative numbers and are an advanced math idea. Just like we can count UP from zero, we can also count DOWN.

That's right! When we move the other direction on a number line, numbers become negative. These numbers are very abstract so they can be hard to think about.

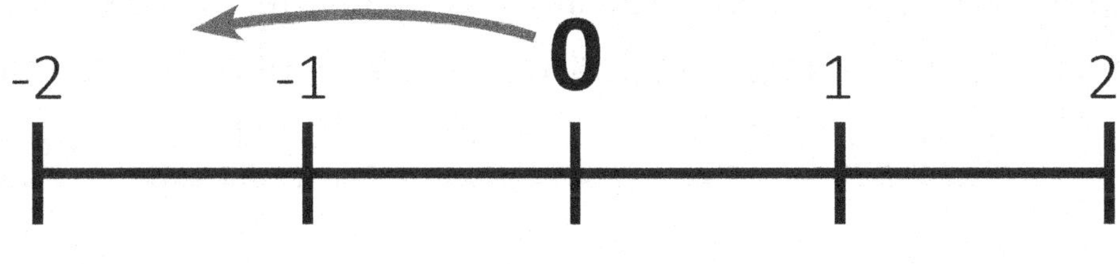

Negatives Visualized

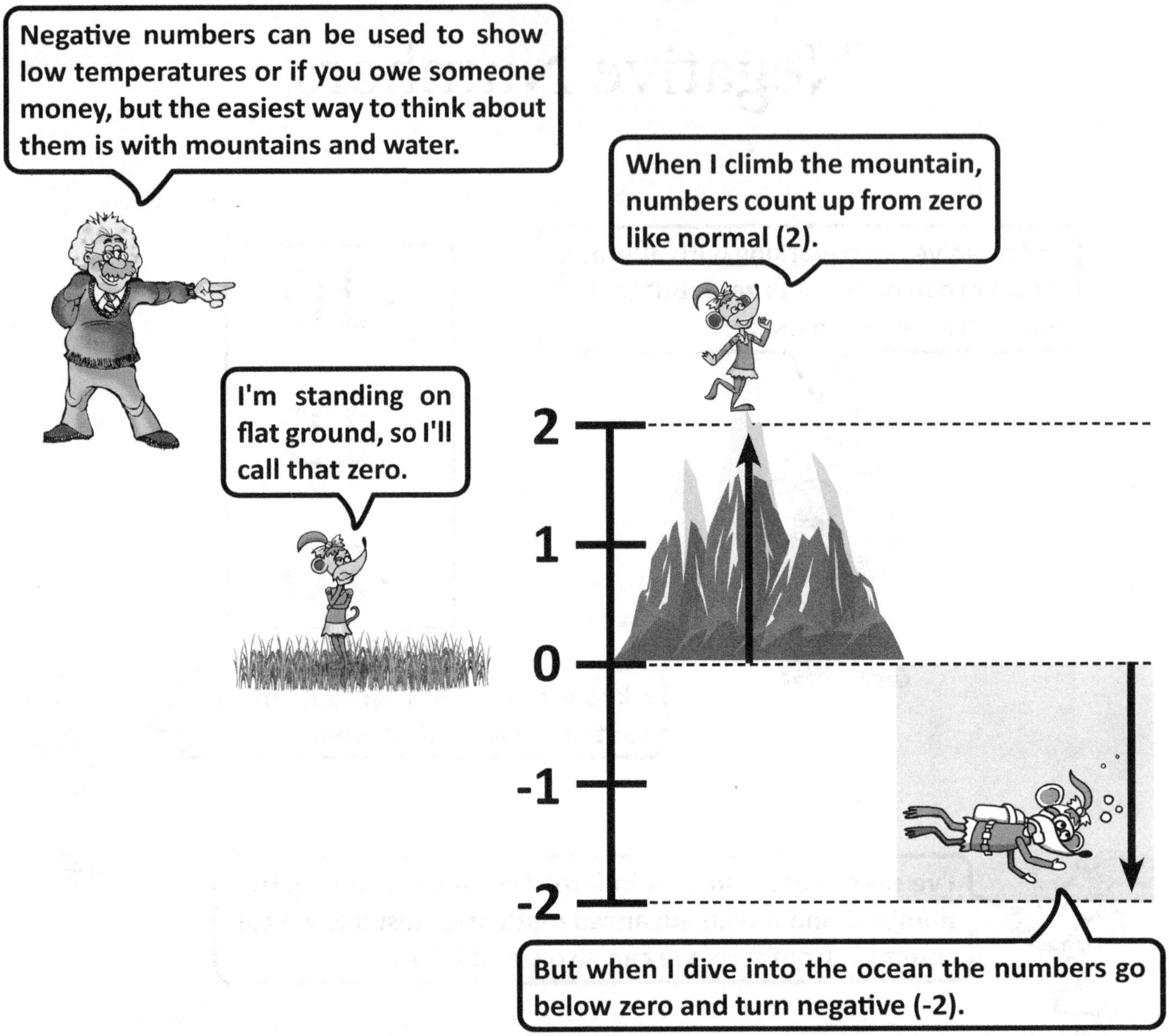

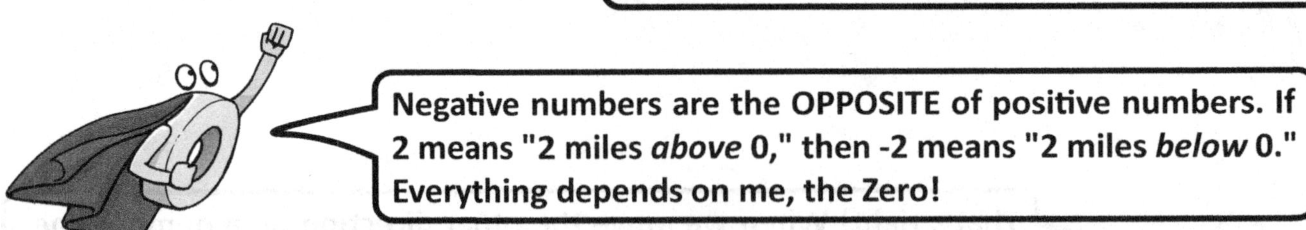

Scuba divers can go 100 feet under water, but to go *really* deep you need a special kind of submarine. The Titanic was found by an unmanned submersible called the Argo. Scuba divers couldn't look for the Titanic because it was on the ocean floor two miles below the surface!

Debt

Negative numbers also apply to money. The opposite of *having* money is *owing* money:

$10 means you *have* $10
-$10 means you *owe* $10

Money in the Bank

It looks like Daniel has the biggest number ($100), but now I see the negative symbol. That means Daniel is in trouble because he *owes* the bank $100.

If I think back to the underwater example, that means he is "in the hole" and has less than zero dollars!

That means that even if I mow lawns all summer and make $80, I will STILL have less than zero dollars. I would need to make $20 more just to climb out of my hole and get back to zero.

Combining Positives and Negatives

Positive and negative numbers can also be combined! Think about a mouse climbing ladders above and below ground (zero) for these examples:

-3 + 3 = 0	-2 + 5 = 3	-3 - 2 = -5

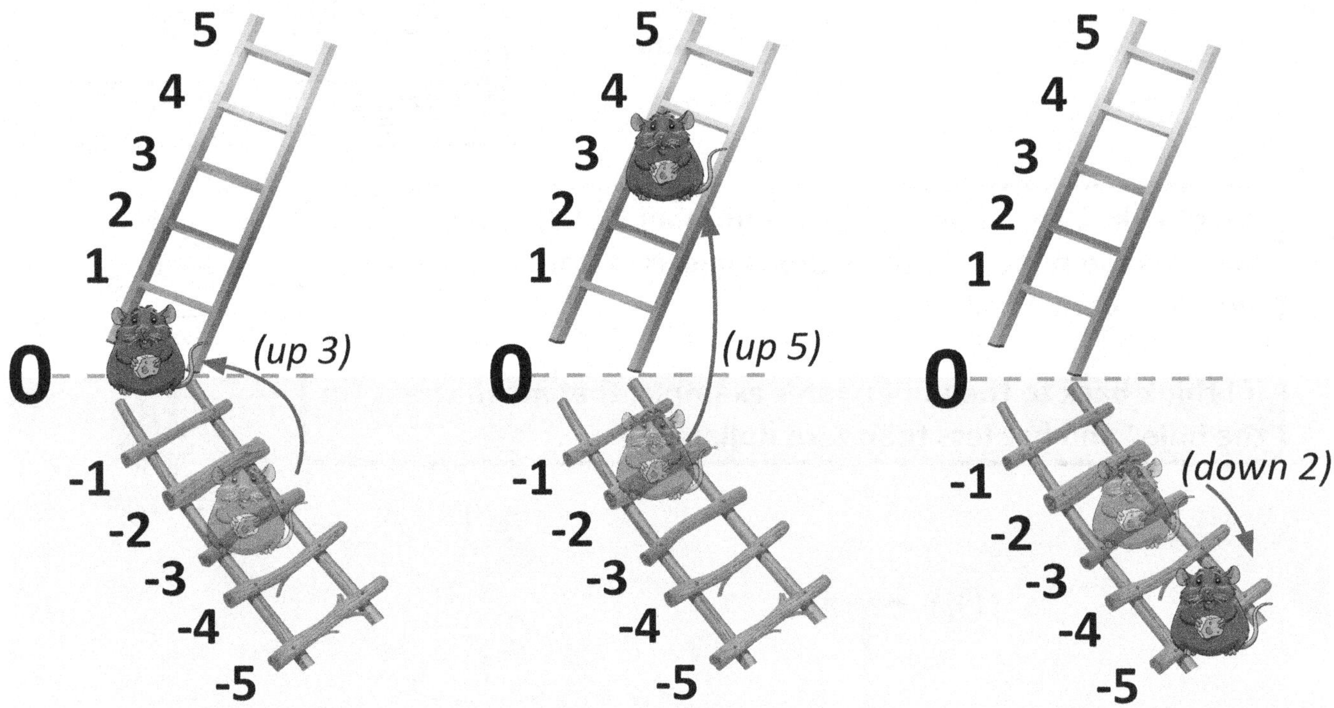

Negative numbers make sense when diving into oceans and climbing ladders, but how can a *temperature* be negative?

It all depends on what we call 0 degrees. Most countries use "when water freezes" as their baseline of zero.

So any temperature above zero (positive) will be warmer than this freezing point, and anything below zero (negative) will be colder.

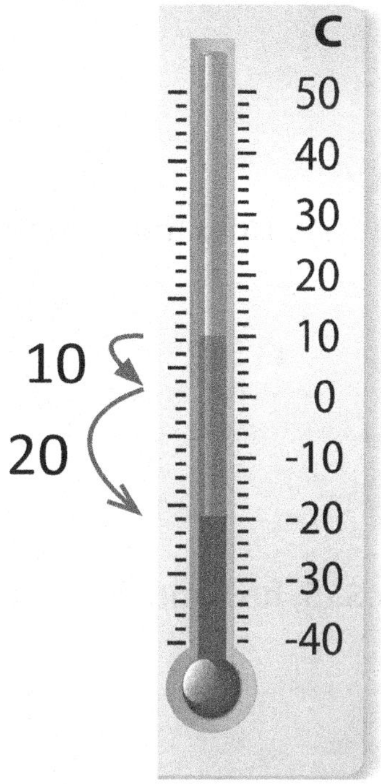

So if the temperature drops from 10° to -20°...

It drops from 10° to 0° (goes down 10°), but then it drops from 0 to -20° (goes down 20° more). That means it dropped a total of 30 degrees!

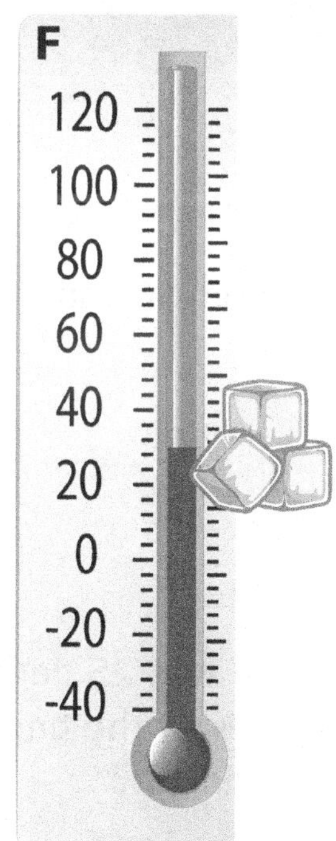

In America, we use a different scale for temperatures. Here, we say that water freezes at 32°, so by the time our thermometers get to negative numbers it means the weather is VERY cold.

Remember to use the right language for negative numbers.

-20 is "negative 20," NOT "minus 20"

Using the right language helps us be more clear about what we are asking children.

Negative Numbers

Level 1

1) Stanley has no money in the bank and owes his dad $12. How much money would you say that Stanley has?

a) -$12 b) $0 c) $12

2) If Tom has $10 in his bank account and owes his sister $15, how much money does Tom have?

3) Nancy owes her mom $6 and owes her dad $14. How much money does Nancy have?

4) If the temperature was 25 degrees in the afternoon and then dropped 26 degrees by the time it was midnight, what was the temperature at midnight?

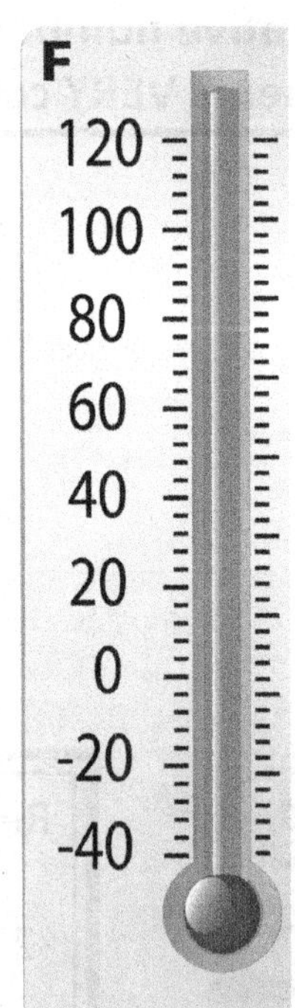

5) If the temperature was -5° in the morning and then dropped 5 more degrees by noon, what was the temperature at noon?

Negative Numbers

Level 2

1) 50 - 40 is equal to 10. What is 40 - 50?

2) Micheal has $25. His friend Dave has -$15. How much more money does Micheal have than Dave?

3) Ed and Sara each have no money and each owe $100 to a bank. Together, how much money do they have?

4) A scuba diver swimming 100 feet below the water's surface is at a depth of -100 feet. A boat captain was sitting on the boat 6 feet above the water's surface. How much higher is the captain than the scuba diver?

a) 100 feet b) 6 feet c) 106 feet

5) Three friends decided to combine all their money. How much do they have altogether?

$20

$30

-$50

Negative Numbers

Level 3

1) 500 - 400 is equal to 100. What is 475 - 550?

2) Carl has $75 and Rachel has -$75. How much more money does Carl have than Rachel?

3) Water was boiled at a temperature of 212°. How many degrees does the temperature have to drop to freeze at 32°?

4) Mt. Everest is about 6 miles tall and The Mariana Trench is about 7 miles deep. How much higher is Mt. Everest than the bottom of The Mariana Trench?

5) Three cats decided to combine all their mice dollars. How much do they have all together?

-$15 -$30 -$150

Negative Numbers

Genius Level

1) The coldest temperature possible is about -460°F. This is called absolute zero and is the temperature where molecules stop moving. How much colder is absolute zero than the temperature where water freezes (32°F)?

2) Mercury is one of the only metals that is liquid at room temperature (about 70°F).
Mercury becomes solid at about -40°F.
Copper becomes liquid at a temperature of 1983°F (and is solid at any lower temperature).
How much lower is -40°F compared to 1983°F?

3) What is -475 - 525?

4) The average temperature at the North Pole is about 32°, and the average temperature at the South Pole is 108° *colder* than that. What is the average temperature at the South Pole?

5) When the wind is blowing, cold weather feels even colder. For example, when the temperature is 10° and the wind is blowing at 40 miles per hour, the temperature feels like -15°.

If the temperature is 10°, how much colder does it feel for someone standing in the wind compared to someone standing without wind?

27
Conclusion

No single book can be a substitute for 10 years of school curriculum. However, any student who works through these chapters will emerge with a solid foundation of numbers that can be applied to any math-related school work (and many real-life problems).

If you take away only *one* thing from this book, have it be the Abstract Dial. This is a universal strategy that will work for every math idea from preschool through 8th grade.

It gives us a starting point to move on from "I don't know," to "maybe if I tried *this*..."

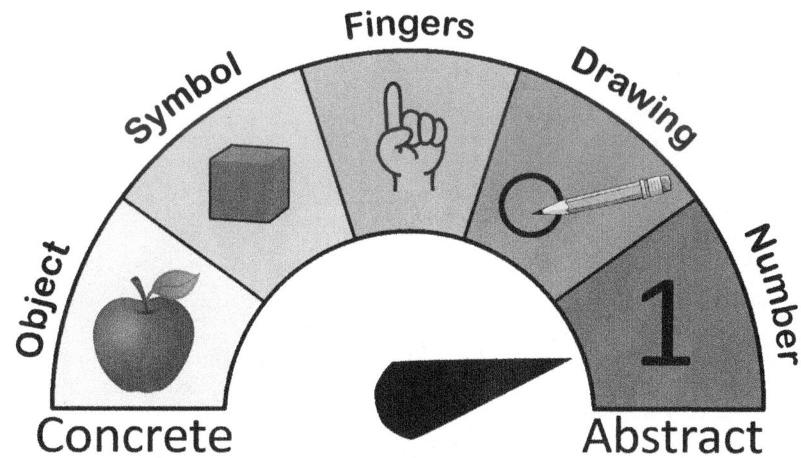

When people learn that numbers follow predictable patterns (and are not random symbols), they will begin to discover the joy of mathematics. Foster that joy in your own mind, and spread it to as many people as you possibly can.

Part VII

Answers

(Section III and Section IV)

Chapter 13 (Page 121): Addition and Subtraction

Level 1

1) Yes, 3 bananas
2) 11 toys
3) 10 pennies
4) 3 gummy bears
5) 3 rocks
6) 9 pennies
7) 22 pieces of candy
8) 31 steps
9) 21 grapes
10) 20 presents

Level 2

1) 7 raisins
2) Any two groups splitting up 8 apples
3) 5 steps (5th step)
4) 7 rocks
5) 10 more pinecones
6) 10 cards
7) 10 circles
8) 13 pieces of candy
9) 40 steps
10) 9 apples

Level 3

1) 46 pieces of candy
2) 47 meals
3) 12 marbles
4) 20 napkins
5) 5 cookies
6) 40 cards
7) 32 pieces
8) 4 cheese wedges
9) 40 shells
10) 6 more tulips

Genius Level

1) 29 squares
2) 43
3) 68
4) 82 legs (remember to include Alice!)
5) 15 more checkers
6) 75 cards
7) 4,660
8) 55
9) $40
10) 45 books

Chapter 14 (Page 135): Place Value

Level 1

1) 13 years old
2) 52 marbles
3) 20 pinecones
4) 444
5) 522 200
6) 99 cows
7) (Any combination that adds up to 80)
8) 90 (or 9 tens)
9) 52 pounds
10) 40 nails, 9 nails

Level 2

2) 5,321
2) $18
3) 8000 (or 8 thousands)
4) 53 cents
5) $416
6) 200 nails, 90 nails, 2 nails
7) 53,711
8) 110 3,752
9) 2 hundreds, 6 tens
10) 99,993

Level 3

1) 649 steps
2) (Any numbers that subtract to equal 64)
3) 575 steps
4) $900
5) 72 degrees
6) 8,000; 600; 40; 2 (pennies)
7) 80,000 + 8,000 + 800 + 80 + 8
8) 3,000,000 + 400,000 + 50,000
9) 10,580 6,000
10) 20,000 5,000 100 (pennies)

Genius Level

1) 3,480 steps
2) 17 ones
3) 4 hundreds, 9 tens
4) 180 degrees
5) 65,100 feet

Chapter 15 (Page 150): Multiplication and Division

Level 1

1) 16 legs
2) 24 toys; Multiply 3 x 8 (or 8 x 3)
3) 13 breaks into 10 and 3.
   (Multiply each part by 4)
   10 x 4 = **40**; 3 x 4 = **12**; **40** + **12** = 52
4) 22 shoes
5) 6 cookies each
   Make a grid and split them down the middle,
   OR pass them out one at a time,
   OR split each cookie in half.
6) 4 x 6 = 24 and 6 x 4 = 24
7) Anything showing 7 groups of 3
   (or 3 groups of 7)
8) 5 x 4 = 20 (or 4 x 5 = 20)
9) 18 arms
10) 4 x 2 = 8 (or 2 x 4 = 8)

Level 2

1) When multiplying, the order doesn't change the answer. 4 x 8 will be the same as 8 x 4.
2) 1088 ÷ **64** = 17   1088 ÷ **17** = 64
3) No, the plates are not equal so you can't multiply them.
4) 4 mini-pizzas each
5) 5 rows of carrots
6) 5 x 7 = 35 and 7 x 5 = 35
7) 4 cartons of eggs
8) 64 squares
9) 24 bishops
10) 110 socks

Level 3

1) 16 (groups of 4)
2) 102 ladybugs
3) 51 marbles
4) 15 drawings
5) 42 tiles
6) Skip count by 8 until you reach 48,
   OR count the rows of 8.
7) 4 pounds
8) 12 days
9) 3,000,009
10) 88 fingers, 22 thumbs, 110 toes

Genius Level

1) Multiple options. One way:
   1000 x 2 = 2000
   200 x 2 = 400   Sum = 2460
   30 x 2 = 60
2) 5 teams
3) 96 chess pieces
4) 6 spiders, 7 grasshoppers, 7 dogs, 6 ducks
5) 132 markers
6) 7, 14, 21, 28, 35, 42, 49, 56, 63, 70, 77
7) 2 x 28, 4x14, and 7x8
8) 3^3
9) 12 rows of crayons (6 in each row)
10) $1 (one dollar)

Chapter 16 (Page 168): Facts: Addition/Subtraction

Level 1

1) 3, 4, 5
2) The amount stays the same
3) +1 is just like counting
4) 4, 2, 8, 12
5) OOOO + OOO = OOO + OOOO (or similar)

Level 2

1)

208 - 3 178 - 170 19 - 17 13 - 1 1,980 - 0

2)

9 + 8 2 + 9 2 + 1 3 + 4 5 + 7

7 + 2 3 + 5 0 + 10 8 + 7 5 + 5

3)

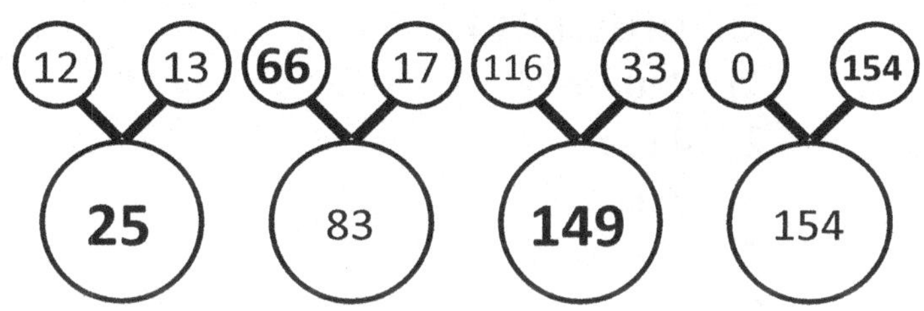

4) x2
5) 13, 12, 12, 24, 31
   (Breaking apart any of the numbers to make a group of 10)

Level 3

1)

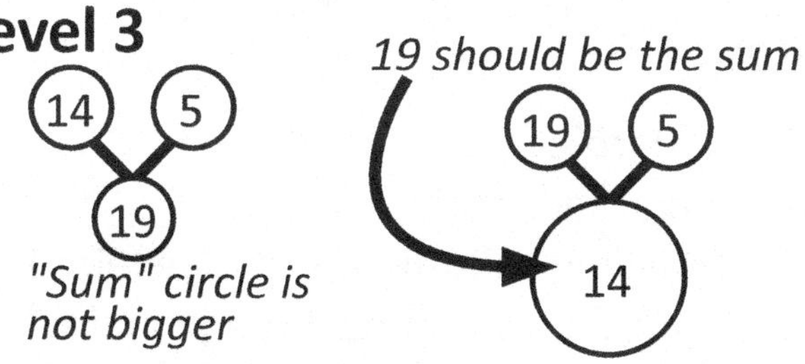

2) 17 11 3 7 12
   9 8 10 15 10
3) Easier: x2 (relates to doubles), x1 (easy to think about "1 group").
   Harder: x0 (abstract), x7 and x8: harder to skip count.
4) $n + 18$
5) 10 is a benchmark number that builds our number system (like place value).

Genius Level

1) $6 - a$, 1 (whole)
2) $2n$ or $n + n$
3) 12, 24, 36, 48, 60, 72, 84
   96, 108, 120, 132, 144
4) 90, 81, 72, 63, 54, 45, 36, 27, 18, 9, 0
5) $n - a$

Chapter 17 (Page 184): Facts: Multiplication and Division

Level 1

1) 12, 6, 2, 18, 14
2) 10, 50, 170, 920, 40
3) 6x**5** is half the number of groups, so the total amount will be half of 6x**10** **(30)**
4) 1 group, or 1 item per group
5) x2

Level 2

1) 32, 20, 0, 24, 12
2) 10, 35, 30, 5, 0
   40, 25, 15, 20, 45
3) 5, 8, 8
   8, 36, 3
4) x4 facts are **twice** as big as x2 facts
5) 4

Level 3

1) 8, 16
2) 4, 49, 36, 1, 16
   81, 9, 0, 25, 64
3) 56, 72, 42, 12, 72
   30, 48, 63, 15, 42
   (Any strategy that adds/subtracts a group)
4) Multiple options: arrays, groups of objects
5) 6 x **4** = 24, 6 x **4** is 24. **24** + **24** = 48

Genius Level

1) 1, 0
2) $\mathbf{c} \div \mathbf{b} = a$
3) $140 \div \mathbf{n} = \mathbf{7}$ OR $140 \div \mathbf{7} = \mathbf{n}$
4) 2
5) 0

Chapter 18 (Page 201): Fractions

Level 1

1) 5 marbles
2) 4 crackers each
3) (Any line splitting the cookie fairly)
4) (Any 2 lines that split the pie fairly)
5) The two cookie pieces are not equal
6) 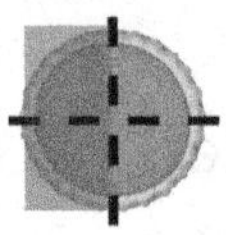
7) ½ pizza
8) Any division that fairly splits the bar (6 squares each)
9) 5 crackers
10) ½ cookie

Level 2

1) 2½ chocolate bars
2) ½ pizza 
3) 1½ cookies
4) 7½ crackers
5) ⅛ pie
6) ¼ pizza
7) They are both right (½ = ²⁄₄)
8) ½
9) ¾
10) Fractions can only be compared if the "whole" is the same. The number of blue and red marbles would need to be the same to compare fractions.

Level 3

1) 6 ½ cookies
2) ½ pizza
3) ⅛ pie
4) ⅝ pie
5) ¼ pie
6) 9⁄10 (the "missing" piece is smaller)
7) ¾
8) ½
9) ¼
10) 1½ pizza 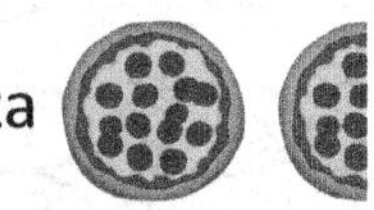

Genius Level

1) 8 jumps
2) 4 pies
3) ⅛ cookie
4) 3¾ cookies
5) $20
6) 7 years old
7) ⅔
8) ¼ pie
9) 128
10) 24 carrots

Chapter 19 (Page 218): Percents

Level 1

1) $4
2) $15
3) 3 eggs
4) 8 chicken nuggets
5) 25%
6) 150
7) 1 pound
8) $4
9) $600
10) $24

Level 2

1) 7% tax
2) $12
3) 6
4) 2 cents or $0.02
5) 20%
6) 100%
7) 32%
8) 86,551
9) 50%
10) $100

Level 3

1) $2
2) 20%
3) 20%
4) $30
5) 200%
6) $11
7) $14.50
8) 50%
9) $25
10) $100

Genius Level

1) 1st year: 150" (12½ feet)
   2nd year: 225" (18¾ feet)
2) 25%
3) 90%
4) 80%
5) 50 pounds
6) 200%
7) 300%
8) 95%
9) 50 cents or $0.50
10) $10.25

Chapter 20 (Page 240): Decimals

Level 1

1) 2 tenths $\frac{2}{10}$
2) 1 bar plus $\frac{2}{10}$
3) 7 hundredths $\frac{7}{100}$
4) (9 sections shaded) **.1 left**
5) 20%
6) .27
7) 

8) .7
9) 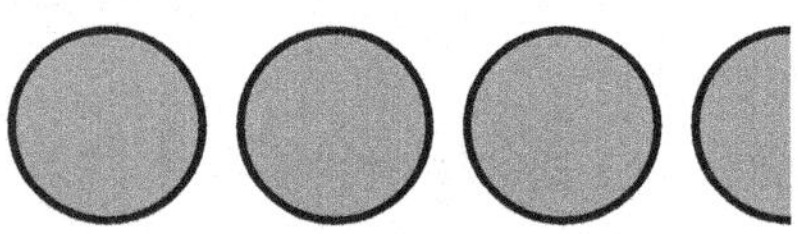
10) .4 or .40

Level 2

1) 45 hundredths $\frac{45}{100}$
2) $\frac{1}{4}$, 25%, 0.25
3) .30
4) .6 or $\frac{6}{10}$
5) $\frac{1}{5}$ and 20%
6) .3
7) 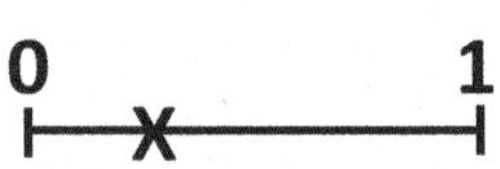

8) .28, .29, $\frac{3}{10}$
9) 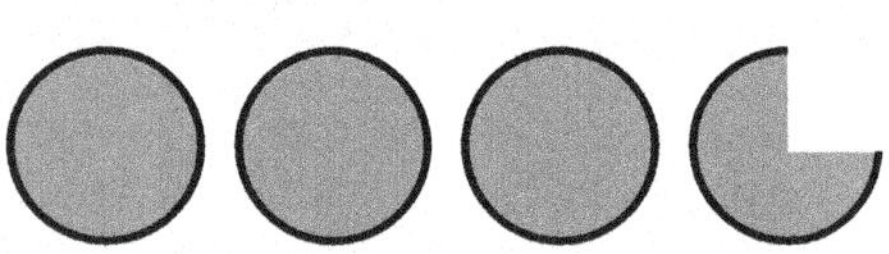
10) 1

Level 3

1) .2
2) 1.1
3) 

4) three hundred, seven five thousandths $\frac{375}{1000}$
5) .03, $\frac{2}{10}$, 21%
6) .2
7) \$3 + 4 dimes + 5 pennies (or equivalent)
8) 50 + 6 + $\frac{1}{100}$ + $\frac{1}{1000}$
9) .08
10) .9901

Genius Level

1) .09
2) $3\frac{1}{3}$ pizzas
3) 1.08
4) 

5) 

6) .027 *(0.28% is equal to 0.0028)*
7) .032, $\frac{3}{10}$, 33%
8) 3.9
9) \$4, 58 cents
10) 70 + 8 + $\frac{1}{10}$ + $\frac{2}{100}$ + $\frac{3}{1000}$ + $\frac{4}{10000}$

Chapter 21 (Page 252): Estimation

Level 1
1) \$5 or \$6
2) \$12
3) \$25
4) 50 times
5) \$7 or \$6.75

Level 2
1) 1000 or 1100
2) 50 rotten apples
3) 120 minutes (15 minutes x 8 miles)
4) 1½
5) 3 hours

Level 3
1) 50,000 (1000 x 50)
2) \$20 (\$2.50 x 8)
3) Almost all of them are red
4) 7 or 8 hours
5) 8 hours (4000 miles ÷ 500)

Genius Level
1) 100 times (600 ÷ 6)
2) 5 seconds
3) 2½ hours (10 miles x .25 hours)
4) d) 1,000,000 times faster
5) c) 8 minutes

Chapter 22 (Page 263): Language of Algebra

Level 1
1) $n + 3$ (years old)
2) $n + 20$ (feet)
3) $n - 10$ (dollars)
4) $n + 15$ (years old)
5) $2n$ (duck legs)

Level 2
1) $60n$ $(60 \times n)$
2) $6n$ $(6 \times n)$
3) $2n$ $(2 \times n)$
4) $8n$ $(8 \times n)$
5) $10n$ cents or $0.1n$ dollars

Level 3
1) $2n + 5$ (pounds)
2) $n/4$ giraffes. (or $n \div 4$)
3) $36n$
4) $4n + 10$
5) $30n$ cents or $0.3n$ dollars

Genius Level
1) $12n + 12z + 6y$
2) $n/60$ (hours) or $n \div 60$
3) $100 - n$ (cows)
4) rat: $2n$ Samantha: $6n$
5) $35n$ cents or $0.35n$ dollars

Chapter 23 (Page 272): Balancing Equations

Level 1

1) $n = 13$
2) $n = 12$
3) $n = 100$
4) 5 bricks
5) $n = 60$
6) $n = 0$
7) $n = 0$
8) $n = 1$
9) $n = 6$
10) $n = 0$

Level 2

1) $n = 9$
2) $n = 5$
3) $n = 90$
4) $n = 0$
5) $n = 1$
6) $n = 5$
7) $n = 2$
8) $n = 25$
9) $n = 25$
10) $n = 12$

Level 3

1) $n = \$1$
2) $n = 31$
3) $n = .5$ or ½
4) $n = 5$
5) $n = .5$ or ½ (or 50 cents)
6) $n = ½$
7) $n = 2$ cents
8) $n = 1000$
9) $n = 45$
10) $n = .25$ or ¼

Genius Level

1) $n = .25$ or ¼
2) $n = .875$ or ⅞
3) $n = 6$
4) $^{96}/_{8} = n$ or $96 \div 8 = n$ ($n = 12$ spiders)
5) $2n + 8 = 40$ ($n = 16$)
6) 86 cows in each group
7) $n = .5$ or ½
8) $n = 12$
9) $2n + 4 = 36$ ($n = 16$ years old)
10) 33 quarters ($25n + 5n = 990$ cents)

Chapter 24 (Page 284): Ratios

Level 1

1) 10 feet
2) 50 feet
3) 10 feet
4) $13
5) $11
6) 2 cups flour
7) 4 feet 3 inches
8) 1 week
9) 300 miles
10) 30 feet

Level 2

1) 1 cup flour
2) 6 feet
3) $8
4) 500 marbles
5) 4 weeks
6) 100 miles
7) 250 pounds
8) 20 feet
9) 12 eggs, 3 pounds of flour
10) 2 pounds silver

Level 3

1) 12 inches
2) 10 cups sugar
3) $2.50
4) 50 feet
5) 3 kiwis
6) 9 pounds
7) 600 owl, 400 eagle
8) 25 weeks
9) 50 feet
10) 2.5 inches or 2½ inches

Genius Level

1) 400 feet
2) 1 quart
3) 5 feet
4)

Mix	Water
0	0
½	¼
3	1½
⅘	⅖
2	1

5) Number lines represent amounts through length, but tables do not.
6) ⅕ ⅖ ⅗ ⅘
   OR 5/25 10/25 15/25 20/25
   OR any equivalent decimals/percents
7) 3 days
8) 3,500 miles
9) 450 feet
10) 50 feet

Chapter 25 (Page 298): Speed, Distance, and Time

Level 1

1) 5 miles
2) 2 miles per hour
3) 25 miles
4) 100 miles
5) 500 miles

Level 2

1) 24 miles
2) 3½ hours
3) 12 miles per hour
4) 11 hours
5) 20 hours

Level 3

1) 25 miles
2) 72 seconds
3) 1 miles per hour
4) 50 miles
5) c) a little more than one second

Genius Level

1) b) About 3 miles
2) 18,000 miles per hour
3) 480 feet
4) b) About 8 minutes
5) 36 minutes (2160 seconds ÷ 60)

Chapter 26 (Page 308): Negative Numbers

Level 1

1) a) -$12
2) -$5
3) -$20
4) -1 degree
5) -10 degrees

Level 2

1) -10
2) $40
3) -$200
4) c) 106 feet
5) $0

Level 3

1) -75
2) $150
3) 180 degrees
4) 13 miles
5) -$195

Genius Level

1) 492 degrees
2) 2023 degrees
3) -1000
4) -76 degrees
5) 25 degrees colder

Part VII

Resources

Scan, copy, or print the following resources to use with your child or classroom!

Concrete to Abstract Dial

Use this dial as a reminder for how to make numbers more basic (Concrete) or more advanced (Abstract).

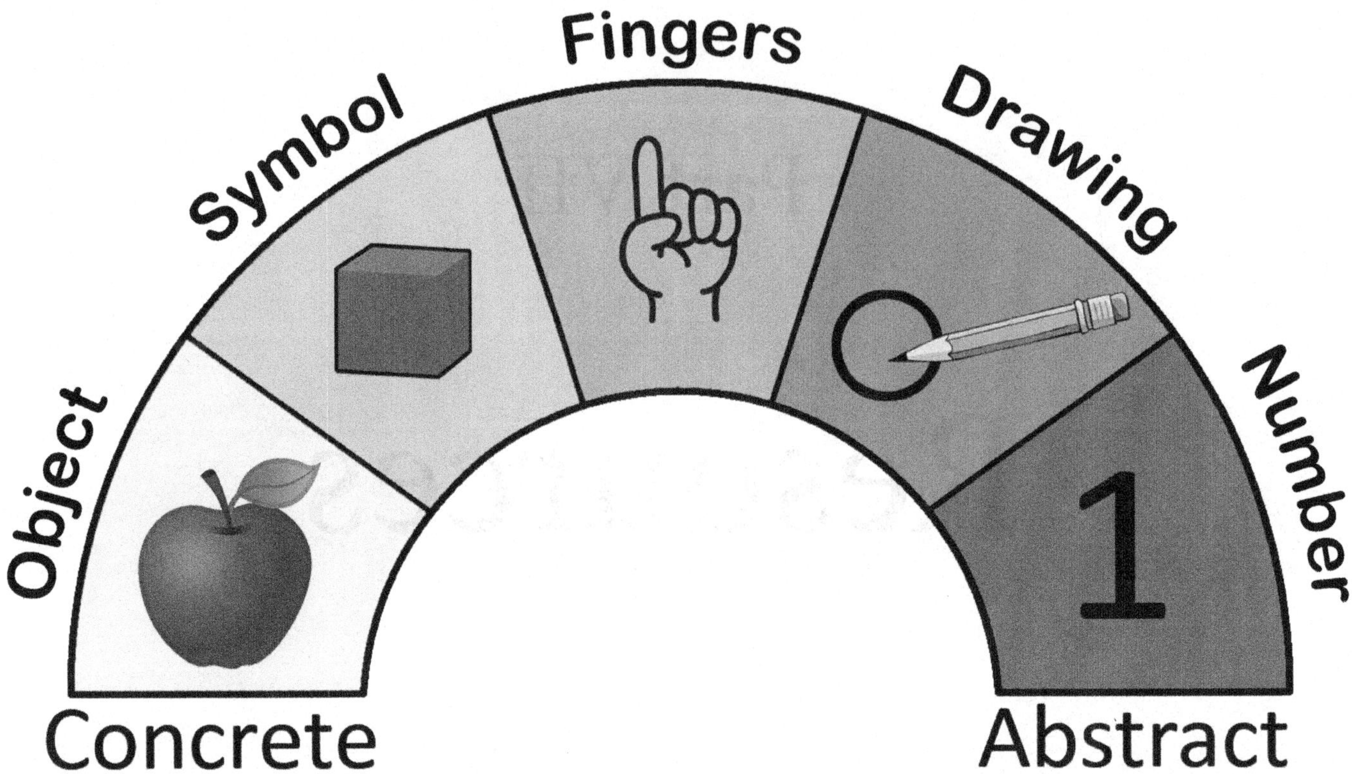

Number Paths / Number Lines

Vertical number lines help show amounts because they get taller as numbers increase. Number paths are an object-based number line. Double number lines can be used to compare ratios or fractions/percents.

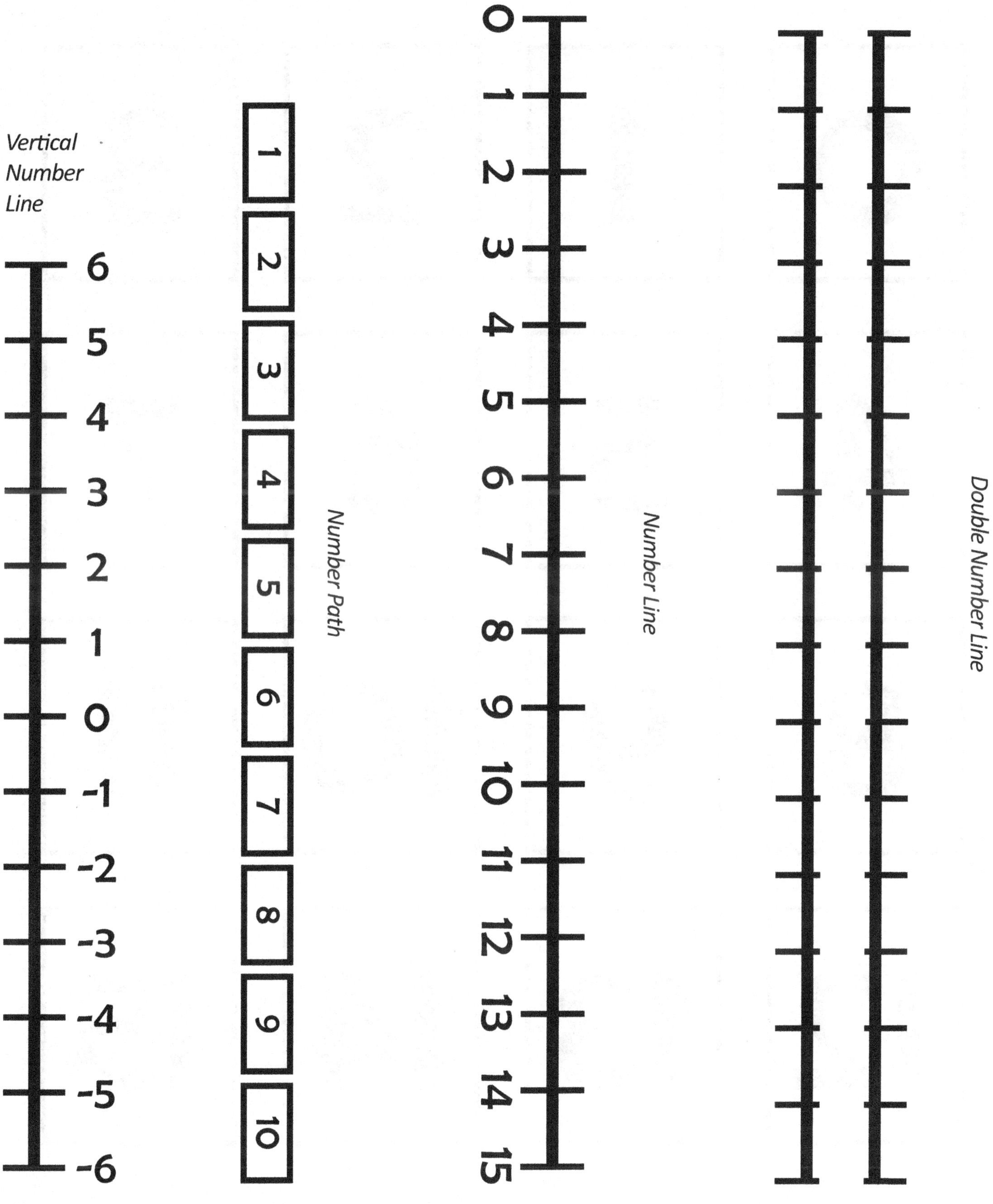

Number Cards

Use these with games like Memory or War, as matching practice with dot cards, or in combination with Place Value cards.

0	1	2	3
4	5	6	7
8	9	10	11
12	13	14	15

Number Cards (continued)

16	17	18	19
20	21	22	23
24	25	26	27
28	29	30	

Dot Cards (Geometric)

Use these to practice recognizing small amounts, for games like Memory or War, or as matching practice with number cards.

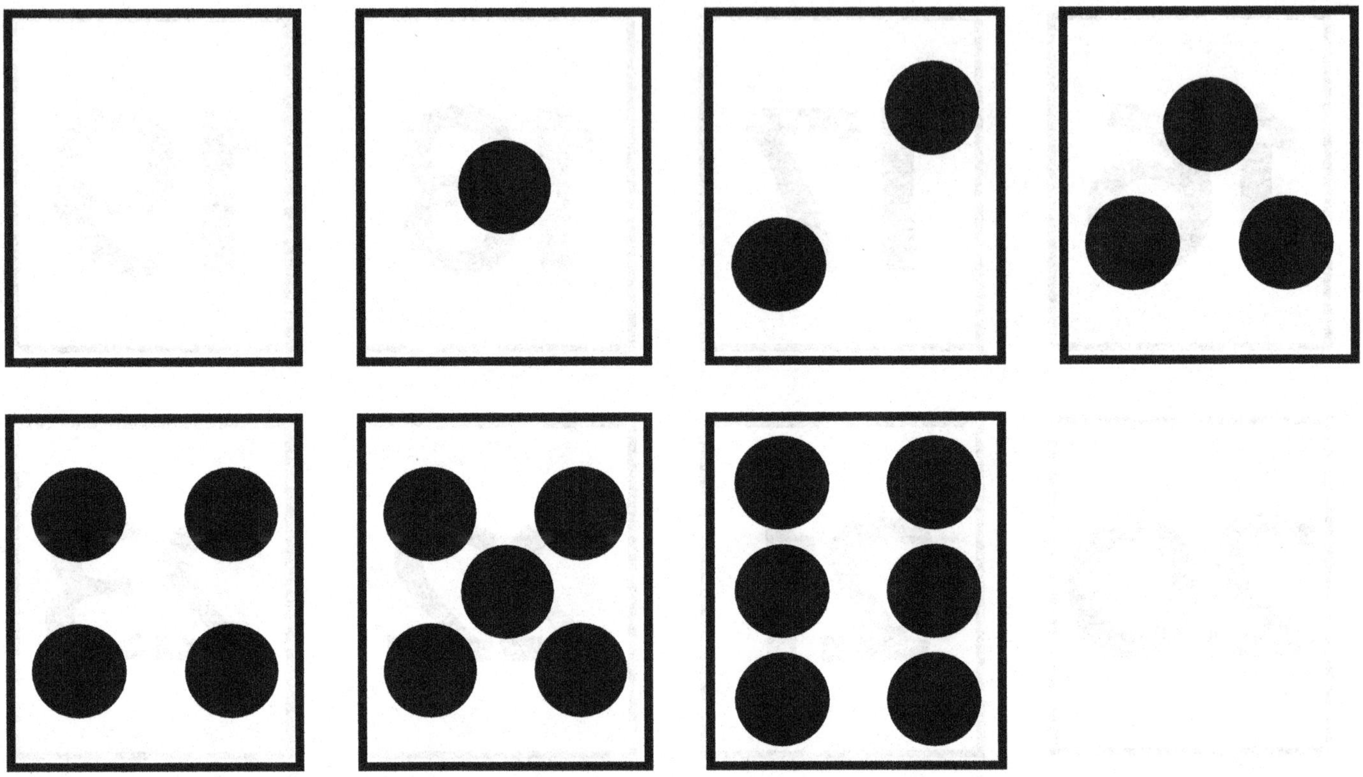

Dot Cards (Scattered)

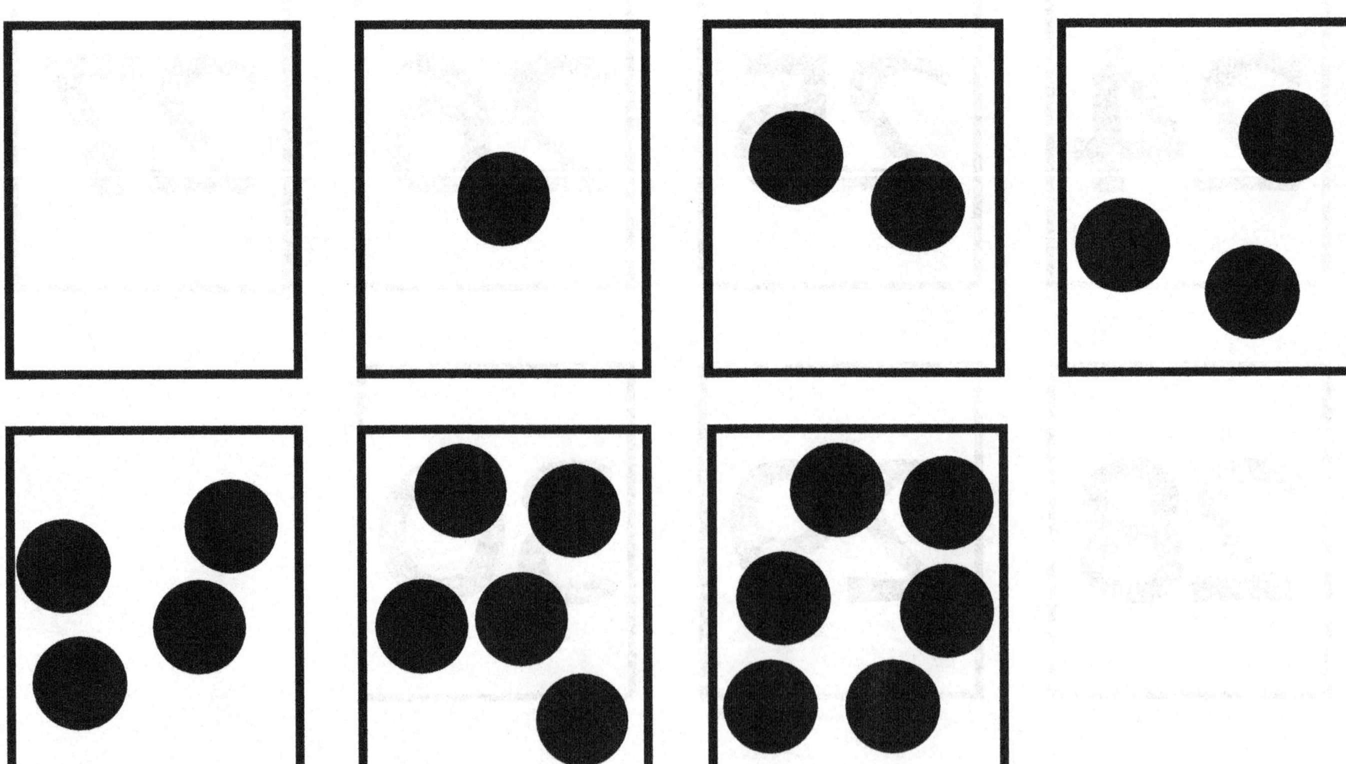

Ten Frames

Use these with pennies or tokens to show ideas like "combinations of 10," "10 and __", recognizing small amounts, joining/separating numbers, and more!

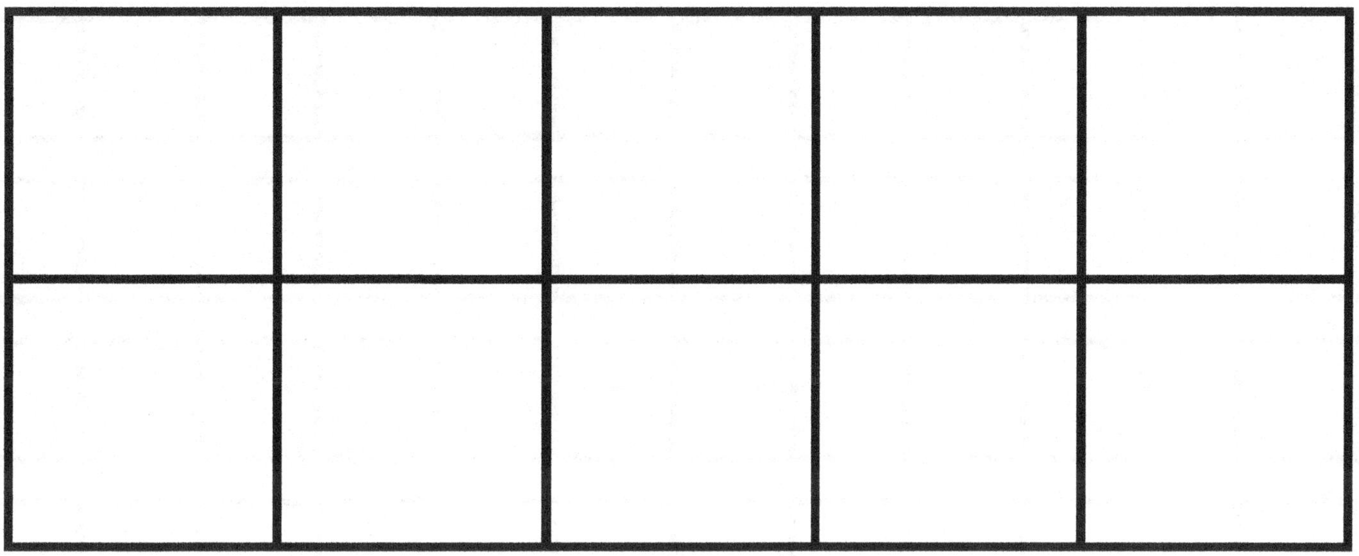

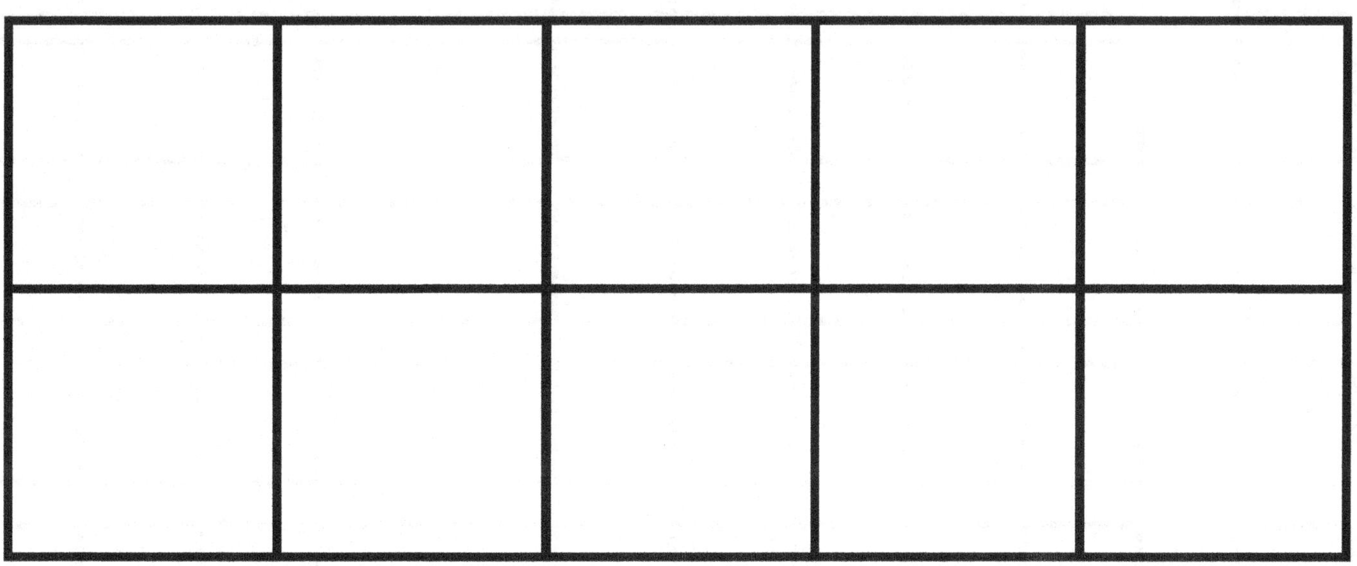

Place Value Blocks

Physical blocks work better, but paper can be used if needed.
NOTE: Call them "10-strips," NOT longs or skinnies.

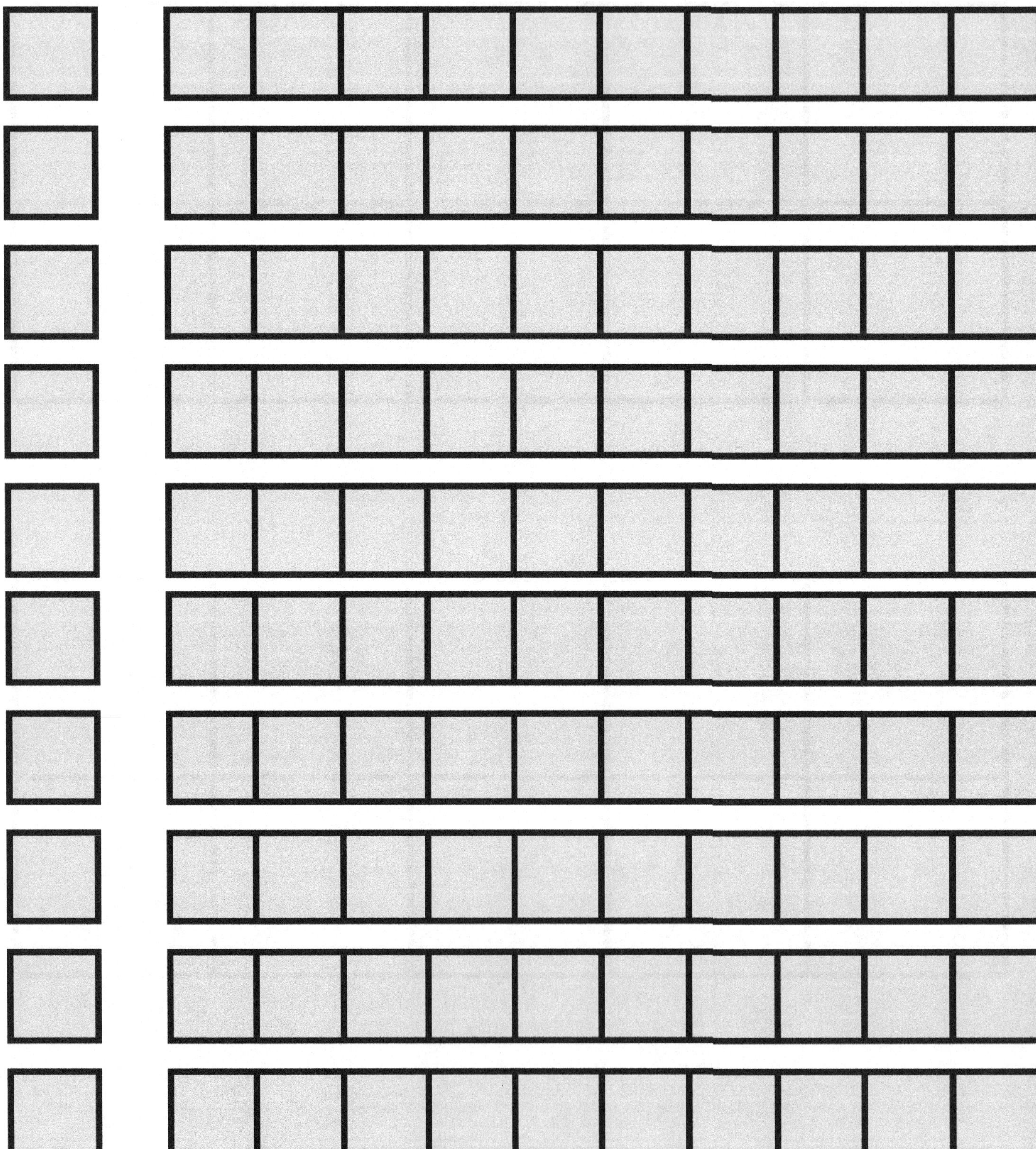

Place Value Cards

These help show the "hidden numbers" of Place Value. Stack them with the single-digit Number Cards.

1 0	2 0
3 0	4 0
5 0	6 0
7 0	8 0
9 0	1 0 0

Place Value Cards (continued)

900	600	400	200
800	700	500	300

Basic Place Value Mat

This helps translate numbers into their place value parts. Numbers can be written in each column to represent the value, or you can use place value blocks.

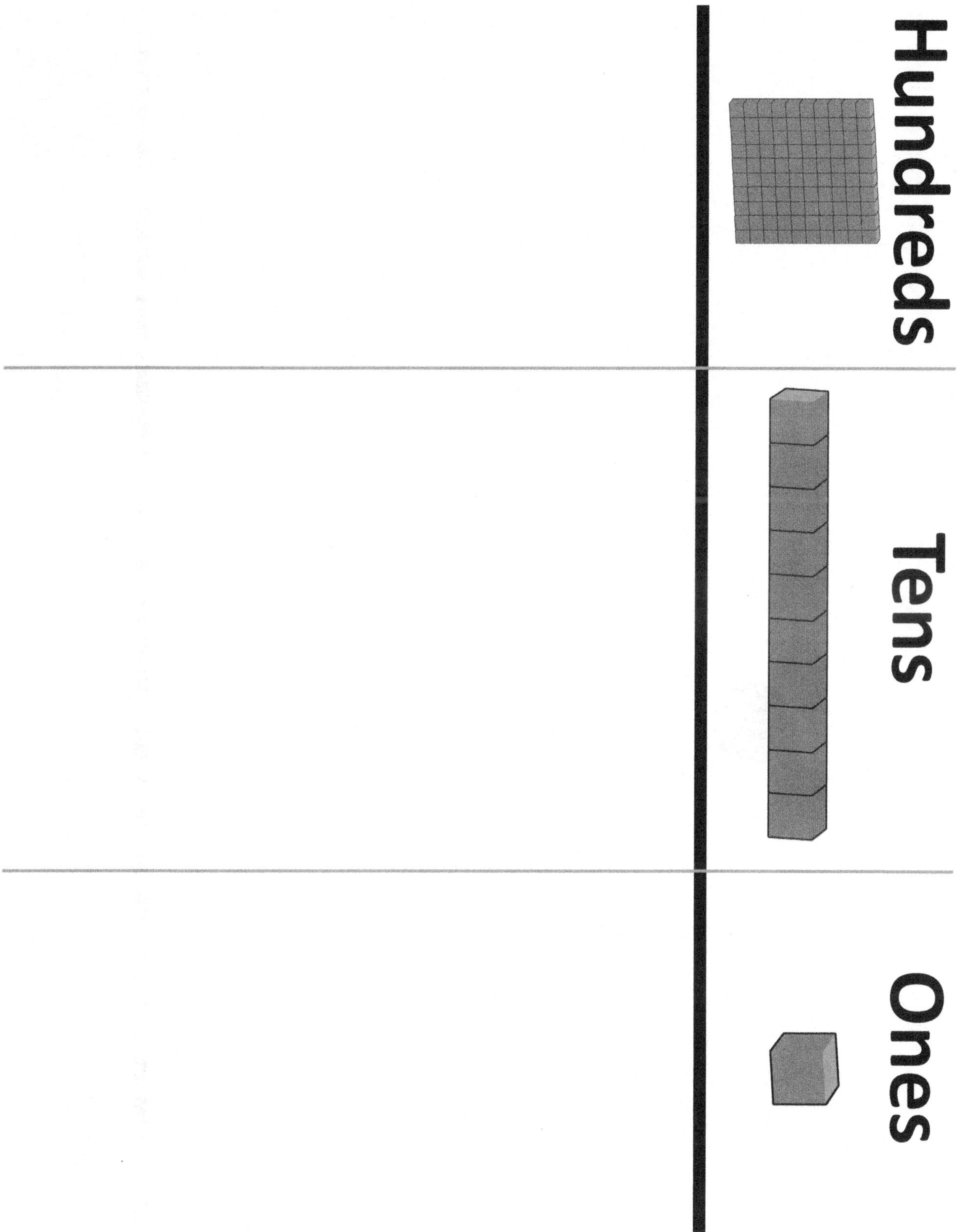

Advanced Place Value Mat

This place value mat includes decimals.

Bingo Board (Blank)

Fill this out with numbers or equations to practice different categories of math facts.

		FREE		

Fraction Strips

These reinforce the idea of "equal parts" and show how different fractions relate to each other.

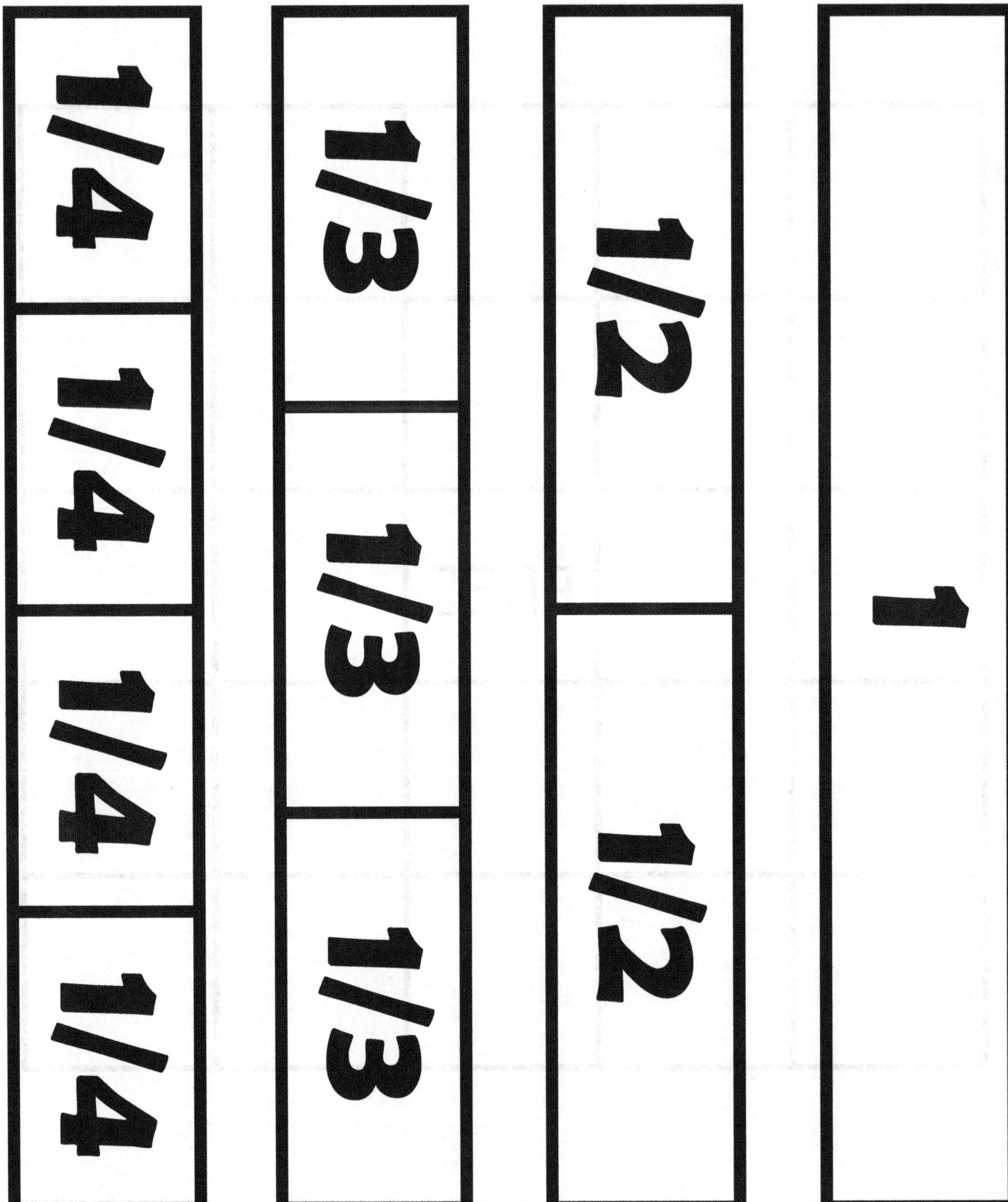

Fraction Strips (continued)

1/8 1/8 1/8 1/8 1/8 1/8 1/8 1/8

1/7 1/7 1/7 1/7 1/7 1/7 1/7

1/6 1/6 1/6 1/6 1/6 1/6

1/5 1/5 1/5 1/5 1/5

References

13 Rules that expire (Karen Karp, Sarah Bush, & Barbara Dougherty, 2014) https://www.nctm.org/Publications/Teaching-Children-Mathematics/2014/Vol21/Issue1/tcm2014-08-18a_pdf/

https://www.achievethecore.org

Big Ideas of Early Mathematics: What Teachers of Young Children Need to Know (The Early Math Collaborative- Erikson Institute, 2013)

https://www.buildmathminds.com/

Cognition-Based Assessment & Teaching of Place Value: Building on Students' Reasoning (Michael Battista, 2012)

Developing Math Concepts in Pre-Kindergarten (Kathy Richardson, 2008)

https://www.illustrativemathematics.org

Math Fact Fluency: 60+ Games and Assessment Tools to Support Learning and Retention (Jennifer Bay-Williams & Gina Kling, 2019)

Made in the USA
Monee, IL
18 October 2026

66036809R00188